CHINA: 7000 YEARS OF DISCOVERY

China's Ancient Technology

Written and Edited by:
China Science and Technology Museum and China Reconstructs

Published by:
CHINA RECONSTRUCTS Magazine, Beijing, China, 1983

ISBN 0-8351-1205-5

Foreword

China's achievements in science and technology in ancient and medieval times were no less remarkable than those produced by other civilizations. In some respects China stood at the forefront of the civilized world. Many of the details of these achievements, however, were never recorded. And, as practically all of China's books were written by scholar-bureaucrats who despised the laboring people, even when technological achievements were described, the accounts were overly simple and often contained errors. For this reason, materials provided by archaeological excavations, including source material on the history of ancient Chinese science and technology, are often more reliable and detailed than those from literary sources.

Many of the successful studies made in the past 30 years on the history of ancient China's science and technology are closely linked with the new archaeological discoveries of the same period. The material obtained from these discoveries has provided reliable data for studies on the history of China's technology and promoted research in such fields as metallurgy, ceramics, textiles, architecture and paper making. New discoveries of oracle bones, bronzes and wooden slips and book copies on silk or paper have given us written data long believed lost to posterity.

In fact, research into the history of science and technology based on archaeological data has been called "archaeology of science and technology" to distinguish it from "history of science and technology" in the strict sense, which is based on written data. Chinese archaeologists are now cooperating with textual historians and natural scientists in the study of the history of ancient science and technology in China. Referring to this cooperation, Professor N. Sirin, Editor of *Chinese Science,* has said, "I know of no precedent anywhere for the extent to which these trends have developed." Professor Joseph Needham, after publishing his *An Archaeological Tour* in 1958, made several visits to China to collect new archaeological data for his magnum opus *Science and Civilization in China.* Chinese specialists on the history of technology, too, are frequent visitors to archaeological institutes and excavation sites.

Of course, support and information exchange in the sciences are not one-sided. Specialists on the history of science and technology and natural scientists have helped archaeologists solve problems related to materials, sources and manufacturing techniques of ancient artifacts.

Due to insufficient data, however, Chinese archaeologists still have many questions to tackle, some of which will probably only be settled by further archaeological finds. The cooperation we have begun will surely push forward research on ancient Chinese science and technology and contribute to the world's storehouse of knowledge as well as promote greater friendship between China and other countries.

Xia Nai

Vice-president of the Chinese Academy of Social Sciences and Honorary Director of the Institute of Archaeology under the CASC.

Introduction

The China Ancient Traditional Technology Exhibition is now on display in Chicago and will be shown in several other cities in the United States. It is sponsored by the China Science and Technology Museum as one of a number of cultural exchange projects between the Chinese and American peoples.

On behalf of the Chinese people, I want to extend a warm welcome to our American friends visiting the exhibition. I hope they will learn something about China's advances in science and technology over several thousand years and her early contributions to the progress of mankind. Residents of the U.S. of Chinese descent may take particular interest in this display of the cultural heritage of their ancestral homeland.

The exhibition shows only a small part of China's ancient technology, arts and handicrafts, but it includes the cream of her achievements over several millennia. In the preface to his comprehensive study *Science and Civilization in China,* British scientist Joseph Needham states that China's technological discoveries and inventions were "far in advance of contemporary Europe, especially up to the 15th century." Ancient China was able to build a highly developed civilization and maintain the unity of her vast territories and many nationalities precisely because of her cultural cohesion based on a common technology. Visitors to the exhibit will see many ingenious inventions connected with agricultural production, irrigation works, architecture, bridge-building, transportation, navigation and communication.

Pictures and models of very large-scale constructions are also on display, including the famed Great Wall, the Zhaozhou bridge and the gardens of Suzhou. Some remarkable recent archaeological discoveries are also represented, such as the 2,200-year-old Qin dynasty pottery army and the

65-piece set of chime bells from the Warring States period (475-221 B.C.) which can still be played today. All of these testify to the ancient Chinese people's creative achievements in technology and the arts.

China's past contributions to the world's storehouse of knowledge include the compass, paper, printing with movable type and gunpowder. If for several centuries she fell far behind in scientific development, she is now rapidly catching up. She has launched man-made satellites and made breakthroughs in many fields. Particularly in some areas of medical and biological sciences and in crop hybridization, she has recently made important and original contributions to the scientific heritage of mankind. And China's current modernization program is opening up still broader vistas for science and technology.

May all of you enjoy the exhibition, and may the traditional friendship between the Chinese and American people continue to flourish.

Yi-Sheng T.E. Mao
Vice Chairman, China Association for Science and Technology, and Foreign Associate, National Academy of Engineering, U.S.A.

CONTENTS

Cai Lun – Inventor of Paper

Traditional accounts have it that paper was first made by Cai Lun, who lived in the Eastern Han dynasty and died in 121. As inspector in charge of the imperial workshops, he had wide contacts with artisans and workmen all over the country. After carefully summing up the experience he gleaned from them, Cai Lun succeeded in producing paper and presented it to Emperor He Di in A.D. 105. The materials — tree bark, remnants of hemp, linen rags and fishnets were first boiled and pounded into a loose pulp. After water and plant gums had been added, the solution was strained through a fine screen on which the pulp dried to become sheets of paper.

Papermaking flourished in China in the Tang (618-907) and Song (960-1279) dynasties. A great variety of paper for different purposes appeared. Some were identified by the materials they were made of, such as hemp paper, hide paper, bamboo paper. Some took their names from the places of manufacture, such as the Jiajiang paper of Sichuan and Xuan paper of Anhui. Some of the names show their use, such as window paper, scripture paper and wall paper. Some were named for the effect created in processing such as water-ripple, gold and silver and patterned paper.

Through the years, Xuan paper produced in Anhui province has maintained special renown owing to its outstanding suitability for Chinese ink-brush painting and calligraphy. Although it looks like ordinary paper, its smoothness, durability, whiteness and refined quality produces the most satisfactory effect when artists use it to paint and write on. Toward the end of the Tang dynasty, Xuan paper was designated as a tribute to the emperor.

Processes in papermaking as shown in drawing from the *Tiangong Kaiwu*, a comprehensive encyclopedia on traditional technology written by Song Xingchu and first published in 1637.

Cutting bamboo and washing it in a pond.　　Boiling pulp.　　　　Straining the pulpy solution through a fine screen.

Pressing it dry to form a sheet of paper.

Drying the sheets on a heated wall.

At the beginning of the 3rd century, Chinese paper was introduced to Korea and Vietnam and from Korea to Japan. Somewhere around the end of the 7th century, the technique of papermaking reached India, Nepal, Pakistan and Bangladesh.

In the 12th century Spain learned papermaking from the Arabs and in 1150 built Europe's first paper mill. Soon afterward the technique was taken to France, Italy and other European countries.

In 1575 papermaking reached Mexico, and some time later Australia. By the 19th century, Chinese papermaking had become known throughout the world.

Pasting New Year pictures on doors at the time of the Lunar New Year is a centuries-old tradition. Pictures usually come in pairs because the doors of traditional-style Chinese houses have two leaves.

These pictures of Qin Qiong (left) and Weichi Gong (right), both generals of the Tang dynasty, were made in Weifang, Shandong province. Legend has it that once Tai Zong, the second emperor of the dynasty, fell ill and was disturbed by noisy evil spirits. The two generals volunteered to stand guard at the palace gates and the spirits never showed up again. The emperor, concerned that they might be worn out by their constant vigil, ordered their portraits painted on the gates instead. Presumably he reckoned that the evil spirits would not notice the difference.

From Bamboo Strips to Bound Books

The earliest written records in China date back 3,500 years. These are the oracle bone inscriptions, the accounts of divination incised on flat animal bones by order of the rulers of the later Shang dynasty (16th-11th c. B.C.). Around the 13th century B.C., there appeared a great variety of elegantly fashioned bronze ware. Besides using these as wine and food vessels and weapons, royalty and aristocrats had important events and documents inscribed on them to keep for posterity. Qin Shi Huang, the first emperor to unite China under one rule in 221 B.C. while travelling around the country ordered local officials to have his feats and achievements inscribed on stone.

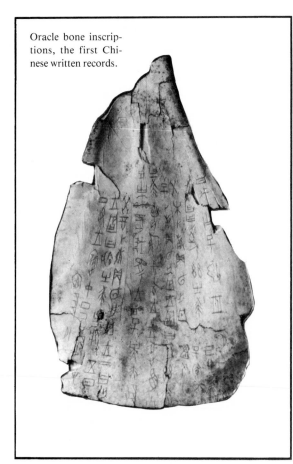

Oracle bone inscriptions, the first Chinese written records.

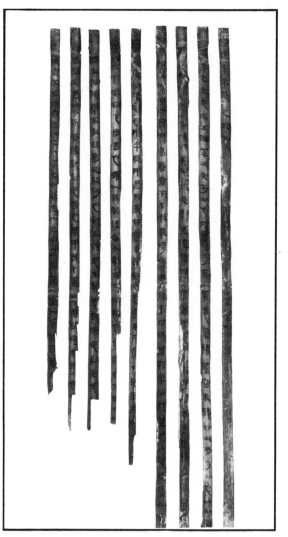

Bamboo strips on which records were written before books came into use.

The earliest books were called *jian ce,* or bundles of bamboo slips. These narrow slips of bamboo on which records and documents were written with ink and brush were strung together with silk cords or strips of cowhide. A single volume was usually made up of a great number of slips. Wooden slips were also used, chiefly for government documents, announcements and correspondence.

During the Warring States period (475-221 B.C.) the Chinese book took a new form. The silk book, being made of soft, pliable material, was rolled instead of being strung together like the bamboo slips. Though never widely used because of the high cost of silk, its introduction spurred the search for a lighter material.

The earliest form of paper book was similar to the silk book. Single sheets were pasted together to form a long scroll. A thin wooden rod was attached to the left side of the scroll so that the paper could be wound around it. Some books were made up of as many as five to ten scrolls.

Since in the long scroll it was not easy to find particular passages, people tried folding it in alternating directions. Thus the long scroll became a rectangular stack and the end-pages were protected with thick paper. The scripture-fold binding

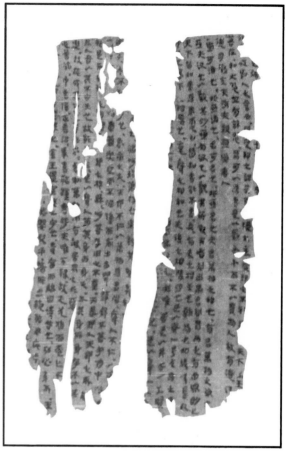

Some early books were written on silk.

Stone table inscriptions.

evolved around the ninth century A.D. Later single pieces of thick paper were used for both front and back covers, joining the two end-pages so that the folds would not fall apart. Then the printed sheet was folded evenly so that the typed pages faced each other. The folded edges were pasted to a strip of paper to form a spine known as butterfly binding, the first prototype of single page binding.

Thread binding came into use during the early 16th century to become the most developed form of the ancient Chinese book. Volumes bound this way have neatly-cut pages of uniform size and are stitched with glossy silk thread. Soft silk material with a beautiful sheen is sometimes used to decorate the covers. Strips of colored silk used to protect the corners lend handsome decoration. After 3,000 years of development, this old method of binding has become a work of art. It is still being used today, especially for new editions of old books.

Most artists of the traditional school use Xuan paper for ink-brush paintings or calligraphy.

Earlier Papermaking Attempts

Three pieces of paper were excavated in December 1978 from the site of a Western Han dynasty (206 B.C.-A.D. 24) building in the Taibai commune of Shaanxi province's Fufeng county. They are thought to date from 91-49 B.C., much earlier than the traditionally assigned date of A.D. 105 for Cai Lun.

The three pieces were preserved in three bronze bulbs decorating a piece of lacquerware. The largest measures 6.8 x 7.2 centimeters, and the thicknesses are 0.022-0.024 millimeters. Made from hemp, they have become mellow, but not brittle. They still retained some luster, and were undamaged by worms or decay.

This is the fourth discovery of hemp paper from the Western Han dynasty. Other samples were found at Baqiao near Xi'an, also in Shaanxi province, in 1957, at Juyan in Gansu province in 1974, and Lop Nor in Xinjiang in 1933. The earliest is that from Baqiao which dates from the time of Emperor Wu Di (140-87 B.C.). The latest Fufeng county finds, though more recent in date, are coarser than the Baqiao samples. They consist of loosely-woven hemp fibres with pulp spread unevenly over them.

Thread-bound books.

Picture on the first page of the *Diamond Sutra* printed in 869, the 9th year of the reign of Xian Tong of the Tang dynasty.

Paper money was used in China as early as 1005 (2nd year of the Jing De reign of the Northern Song dynasty). This is a copper plate used for printing the paper money called *huizi* at Lin'an (Hangzhou), provisional capital of the Southern Song dynasty in the 12th century. Printed on the bills were the name of the issuing agency, the value of the bill and the reward for the first person to report a counterfeiter.

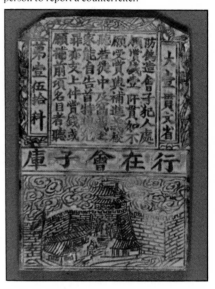

Printing in Ancient China

Before the invention of printing in ancient China, books and property deeds were handwritten. The technique of printing with carved wood blocks appeared about the 7th century, early in the Tang dynasty. Also called block printing, it was done by copying the manuscript according to a set format on thin sheets of paper. These were stuck face down on blocks or hard pear- or date-tree wood and the reversed characters carved in relief. Printing was done by applying a thin coating of ink on the block, after which a sheet of paper was placed over it and stroked with a brush. A printed sheet with the characters right side up resulted. This technique was also used for drawings and illustrations.

Block printing reached its golden age during the Song dynasty (960-1279) as imperial patronage encouraged the publication of large numbers of books by the central and local governments. In A.D. 971, Zhao Kuangyin, the first emperor of the Song dynasty ordered the compilation of the Buddhist scriptures. This work, the *Tripitaka*, appeared in 5,048 volumes. The carving of its 130,000 blocks was done in Chengdu, Sichuan province and took twelve years to complete. Copies were presented by the government to monasteries in the empire and to friendly neighbor states.

A major area of printing had to do with paper money, called *jiaozi* (receipts) when first put out early in the 11th century. This was the first paper money issued anywhere in the world. By the time of the Southern Song dynasty (1127-1279) paper money—now known as *huizi*—was being issued in large quantities. The Huizi mint, located at Lin'an (today's Hangzhou), employed 1,200 workers, of whom more than 200 were printers. The plates, generally of copper were engraved with intricate designs to discourage counterfeiting.

Among later developments in block printing the most characteristic was the multicolor *tao ban* and *dou ban* techniques, which appeared around the time of the Yuan dynasty (1271-1368). For *tao ban* prints, several blocks were made of the same plate, each for a different color, and the impressions from them superimposed on the same sheet of paper. Only two colors were used at first, but later increased to as many as seven.

A more refined variant was the *dou ban* technique (now called woodblock watercolor printing) developed on the basis of the *tao ban* and used mainly for printing reproductions of paintings. Each plate is divided into a number of blocks according to the colors in the design. Then each block is impressed on paper with a different color to produce a complete picture. Colored New Year pictures reproduced in this way became very popular in the 17th century. The best known ones were those made at Taohuawu near Suzhou, and Yangliuqing not far from Tianjin, and Weifang in Shangdong province.

The art shop Rong Bao Zhai is over 300 years old. Reproductions of ancient paintings made here by skilled craftsmen are almost indistinguishable from the originals. Preparing sketches of the original design is the first step in the making of reproductions. Then wood blocks are carved from the sketches, and used for the printing. The final step is mounting the finished product.

Detail from a Rong Bao Zhai reproduction of *Han Xizai's Evening Party* painted by Gu Hongzhong of the Five Dynasties period (907-960).

Yangliuqing New Year Picture—*Guizi Youyu,* on a traditional theme symbolizing prosperity to the younger generation. Yangliuqing, a village near Tianjin in north China, began making such New Year pictures in the Ming dynasty during the reign of Chong Zhen (1628-1644). Scenes from traditional operas and novels as well as beautiful women and chubby children are used to represent happiness, good luck and prosperity.

The Yangliuqing pictures, still today a well-known folk art, are given an added application of watercolor after printing.

Movable type was invented by Bi Sheng of the Song dynasty between the years 1041 and 1048. This is recorded by his contemporary, Shen Kuo, in *Mengxi Bitan* (Dreampool Essays). Bi Sheng fashioned type out of clay and hardened it by firing. Before printing, the type was placed in rows on a mixture of resin, wax and paper ash spread over a framed metal platen. Then the platen was heated to melt the mixture and the type pressed in. Once the platen had cooled it could be used as a printing plate. To release the type for further use the platen needed only to be heated again. This method approximates the basic principles of type-setting in today's printing techniques.

During the 13-14th centuries, the agriculturist Wang Zhen made an important contribution to the development of movable type printing. In the year 1298 he made more than 30,000 pieces of movable type out of wood, using them to print the 60,000-character 100-volume *Jingde Xian Zhi* (Records of Jingde County). The printing took him less than a month. The characters were made of equal size and height, and no glueing substance was needed. Wang Zhen also invented the use of rotating trays on which the pieces were arranged. The typesetter, seated between two such trays, turned them to find the characters he wanted. This made retrieval of the type fast and easy.

Wang Zhen's chapter "Making and Printing with Movable Type" appended to his *Book of Agriculture* is the world's earliest treatise on the subject.

Movable metal type printing began in China in the 13th century with the use of type made of tin. After that copper type became popular. In 1726 the Qing government published the largest work ever printed with movable copper type—the Encyclopedia of Ancient and Modern Books, consisting of 5,200 volumes in 64 sections.

Printing techniques invented in China spread to neighboring Japan and Korea, and later westward to Persia. Their influence was felt in Egypt and Europe.

Rong Bao Zhai Shop

The Rong Bao Zhai Studio, an art shop on Liulichang Street in the southern part of Beijing, is famous for block-printed reproductions of Chinese traditional paintings so well done that even experts sometimes mistake them for the originals. Laymen looking for the marks of photoengraving cannot find them. A famous art expert once declared a reproduction of a Tang dynasty scroll as the original. Once a reproduction and its original were placed before the painter himself, the famous Qi Baishi (Chi Pai-shih). The old man hesitated before he could identify his own.

In its front shop Rong Bao Zhai sells paintings, reproductions, prints and painting materials. Behind it are four large workrooms, each devoted to once of the four processes involved in making a reproduction — tracing, carving, impressing and mounting.

A qualified craftsman must know more than just carving and impressing. For example, to make a reproduction of a Qi Baishi painting of a frog, first of all he has to understand how the painter used the different tones of ink to create the effect of three dimensions and the elasticity of the animal's flesh. He needs to know that the painter did the leaves with a saturated brush to create life and reality. He must know how Qi Baishi's brushwork brought out the hardness of a crab's shell and the softness of the tiny hairs on its legs.

Rong Bao Zhai possesses a reproduction of a famous Tang dynasty painting, *Han Xizai's Evening Party,* a scroll 332.5 x 29.5 cm. by Gu Hongzhong. Done with highly refined craftsmanship, it took two carvers and a worker-artist eight years to finish. The original scroll contains 45 figures in five scenes, the head of each figure no larger in size than a fingernail. Five thousand impressions using 1,600 wooden blocks were required to make the reproduction.

Since the founding of new China the studio has reproduced a great number of ancient and modern works. Its excellent reproductions make Chinese traditional paintings more easily available for specialists to study and at the same time satisfy the needs of both domestic and foreign art lovers.

The main processes of Rong Bao Zhai's woodblock printing technique.

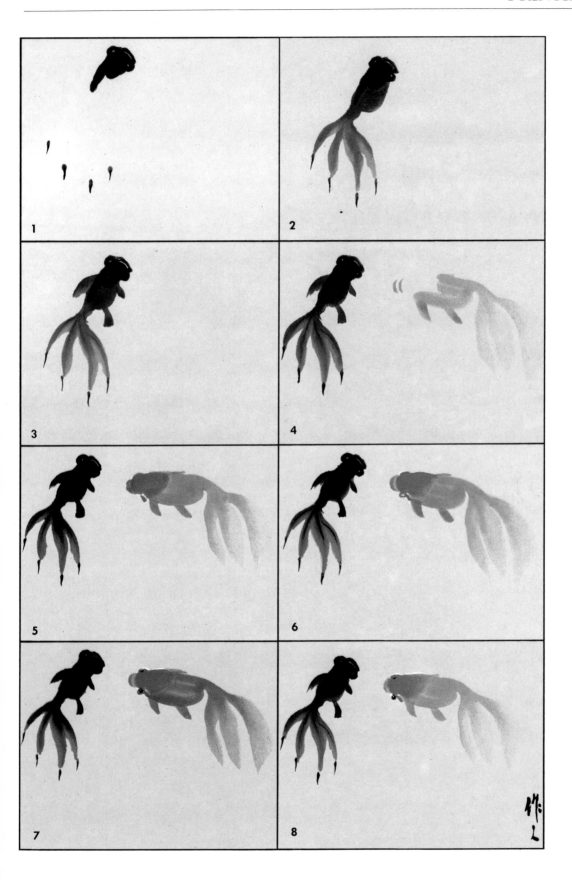

Gunpowder

China's invention of gunpowder was linked with alchemy; it was an experiment for longevity pills that produced the first successful formula.

From the Warring States period (475-221 B.C.) on there had been great interest in eternal life. Taoist alchemists were brought to the imperial court to prepare drugs with cinnabar (mercuric sulphide), arsenic and other minerals thought to be essential ingredients of formulas for immortality.

During the reign of Emperor Wu Di (156-87 B.C.) of the Han dynasty extensive researches for elixirs of life were carried on. Among the substances Chinese alchemists worked with were sulphur and saltpeter, and many fires were started. Such experiments did not yield the desired results but in the course of them discoveries of scientific value were made. The earliest book about alchemy extant today, *Zhou Yi Can Tong Qi* (Book of the Kinship of the Three) by the well-known 3rd-century alchemist Wei Boyang, contains much important information.

By the 8th century in the mid-Tang dynasty, the potentialities of sulphur and saltpeter when combined with charcoal were realized as the alchemists discovered an explosive mixture. This was gunpowder, or *huoyao* (fire medicine) as Chinese still call it.

Sun Simiao, a noted pharmacologist of the Tang dynasty (618-907), in his *Dan Jing Nei Fu Liuhuang Fa* wrote the following prescription for making immortality pills. "Place two ounces of ground sulphur and two ounces of ground saltpeter in a pan for frying. Ignite three gleditsia pods and throw them in to make the mixture burst into flame. When the flame dies down add three *jin* (1.5 kilogram) of wood and the same quantity of charcoal and fry again. Remove from fire when the

Tiger-head shield. This shield, invented in the 17th century, is described in 'Wu Bei Zhi,' a classic military work by Mao Yuanyi of the Ming dynasty. Made of wood covered with sheepskin, it has a hidden box containing eight "fire arrows." Over the tiger's head is a peephole.

charcoal has been reduced by one third." This was, in fact, the world's first formula for gunpowder.

Gunpowder was called "fire medicine" because its three constituents were used separately as medicines. Sulphur was to treat skin disease (as it still is) while saltpeter was used to dispel fever, treat stomach ailments and to disperse internal accumulation of blood. As recently as the 17th century gunpowder was still classified as a medicine due to its use in treating "ringworm sores, insects, eczema and pestilence."

When the use of gunpowder passed from alchemists to military men still remains unknown. Some historians believe this dates back to the late Tang dynasty.

During the reign of Emperor Ren Zong (1010-1063) of the Song dynasty, Zeng Gongliang wrote a military encyclopedia, *Compendium of the Most Important Military Techniques* (Wujing Zongyao), in which the making and use of gunpowder were detailed.

Under the Northern Song, invention of weap-

Discovery and use of gunpowder.

ons was encouraged, and thus inventors usually presented their ideas to the government. In August of the year 1000 the naval commander Tang Fuxian was cited for using rockets and fire balls. Arrows bearing five ounces of gunpowder near the tip were also put into use at that time. According to historical records, a huge quantity of arrows loaded with gunpowder — 17,000 in one day — were used by Song dynasty generals in the 25-day defence of today's Hubei province against the Jin invaders from the north in 1221.

Rocket propulsion as used by the ancients was essentially the same as today. A paper tube attached to the arrow was packed tightly with powder. The continuous jet of rapidly expanding gases propelled the projectile forward. The first rocket of this kind appeared late in the Southern Song period. But true rockets were used in battle for the first time by Kublai Khan, the first emperor of China's Yuan dynasty (1271-1368), in military expeditions against Japan in 1274 and 1281. A Japanese historical record says they "fell for a time like rain."

Also in the Yuan dynasty appeared the first gun to discharge arrows fired by gunpowder. In 1332 the world's first bronze cannon was made. It is now displayed in the Museum of Chinese History in Beijing.

Further improvements were made to rockets during the Ming dynasty (1368-1644). Tips of different shapes gave them names like flying broadsword, flying spear, flying sword and swallowtail.

The "magic fire flying crow," propelled by burning powder on the shaft, exploded on reaching the target. One of the earliest missiles, called the "heaven-shaking thunder gun," was a ball woven of bamboo with paper pasted over the surface and wings on either side. Inside was a paper tube packed with powder and a fuse. A charge propelled it toward its target and it exploded as soon as the charge burned out. "Fire-dragon emerging from the water," the world's first two-stage rocket, was a 1.6-meter tube of wood or bamboo in the shape of a dragon. The first-stage rocket on the outside propelled the tube forward. When it burned out, the fuse in the dragon's mouth ignited the second stage rocket. These rockets were used in naval warfare.

Powder-propelled firecrackers were invented during the Song dynasty reign of Emperor Xiao Zong (1127-1194), and led to festival fireworks. The "box-lantern" was one of such fireworks. It was a multilayered, many-sided firework about one meter high. Inside it were folded fire-resistant paper illustrations of folk-tales, characters from plays and flowers. When it was hung in a high place and ignited it exploded layer by layer so that the story gradually unfolded in whirling spark trails amidst a sea of light.

Gunpowder and firearms, which were to have a tremendous impact on Western history, came to Europe via trade routes and the westward campaigns of the Mongol armies, finally arriving in the 13th century.

Fireworks over Beijing's Tian An Men Square.

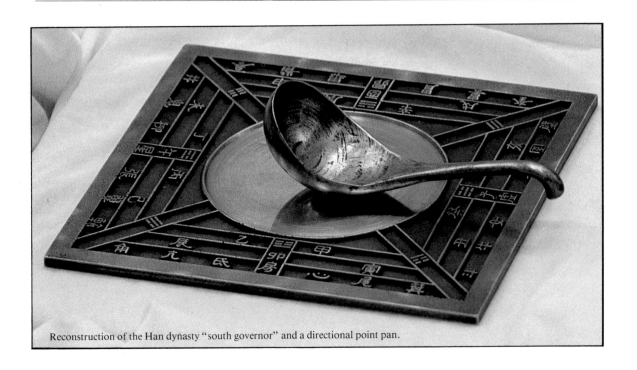

Reconstruction of the Han dynasty "south governor" and a directional point pan.

The Compass

The magnetic north-south pointing properties of the lodestone, or magnetite, may have been used for direction-finding in the third century B.C., or perhaps, if old tales have any validity, even 300 years earlier. The earliest written reference to a south-pointer, which may have been made with lodestone, comes from the third century: "When the people of the State of Zheng go out in search of jade, they carry a south-pointer with them so as not to lose their way in the mountains."

Qin Shi Huang, the First Emperor, who unified China under the Qin dynasty (221-206 B.C.), had the gates of his Afang palace near Xi'an lined with magnetite to prevent anyone entering with concealed iron weapons.

The world's first compass, the "south-governor," was made in China during the Qin dynasty by balancing a piece of loadstone carved in the shape of a ladle on a round, bronze plate. The latter represented the heavens, and was set within a square plate representing the earth. It was engraved with

One of the suspension methods used for ancient Chinese compasses: a magnetic needle hangs from a silk thread.

21

Zheng He (1371-1435), a Moslem from Yunnan province, was a eunuch in the Ming court. By order of the emperor he made seven ocean voyages between 1405 and 1433. His ships, laden with trade goods and gifts of gold, silver, silk, porcelains, iron utensils and cloth sailed to Southeast Asia, Bangladesh, India, Iran and the Arabian Peninsula, going as far as the Red Sea and the eastern coast of Africa. The largest of his ships was 140 meters long, the biggest in the world at that time.

A bowl compass dating back to the 13th century unearthed in 1958 in Lushun, Liaoning province. The inside motif represents a floating needle. The character "needle" is inscribed on the bottom of the bowl.

A Song dynasty compass.

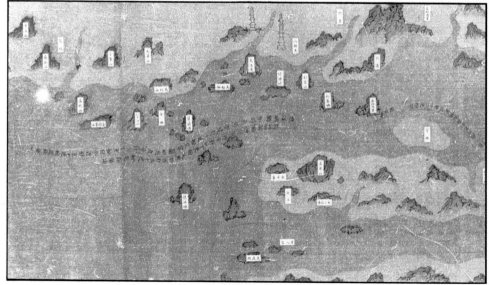

Replica of a chart by Mao Zhiyi of the 17th century which records the seventh voyage by the Ming navigator Zheng He.

24 directional points and, when the ladle was spun, it always came to rest with its handle pointing to the south.

Natural lodestone compasses were still being used in the 10th century A.D., but advances in technology during the 11th century led to the use of artificially-induced magnetism. Two methods of making magnets were known at that time. The first consisted of polarizing pieces of iron by heating them in the earth's magnetic field. Thin strips of iron were cut into the shape of a fish, placed in a fire until they were red hot and then laid down north-south to cool. This magnetized the iron fish so that when they were suspended in bowls of water their heads pointed south. The second method of inducing magnetism was to take an iron needle and rub it with a piece of lodestone. Once an improved method of suspension had been worked out, the magnetic needle proved handier than the fish for a compass.

In *Dream Pool Essays,* a book by the 11th-century scientist Shen Kuo, four ways of suspending a magnetic needle are described: pass it through a piece of rush of the kind used to make lamp-wicks so it can float on water; balance it on a fingernail; balance it on the rim of a bowl; suspend it by a silken thread. With his invention of the silk

Reconstruction of a tortoise-shaped compass described in the book *Shilin Guangji* by Chen Yuanjing of the Southern Song dynasty (1127-1279).

thread suspension Shen Kuo created an instrument so sensitive that he was able to detect the declination between true north and magnetic north 400 years before this was discovered in the West.

The final stage of Chinese compass development was the appearance of the azimuth bearing pan marked with the 24 Chinese compass points and enclosing a bowl of water with a floating compass needle—much like a modern liquid-filled compass. These compasses were used until the 16th century when more convenient dry mountings were introduced.

China's maritime history can be traced back 2,200 years but, before the invention of the compass, it was limited mainly to the coasts. The invention of the compass had an enormous impact upon the art of navigation. With the aid of the compass, seamen gradually explored and established ocean navigation routes which they called "needle routes." This led to a rapid increase in China's foreign trade. By the 13th century A.D. China was trading with more than 50 countries in Africa and Asia.

the voyages of Zheng He in the early 15th century. Between 1405 and 1433 he led seven fleets of huge junks to Southeast Asia, India and the Arabian Peninsula. The greatest of these consisted of more than 60 vessels with a crew of 27,000 men. His largest ship was 140 meters long—the biggest in the world at that time. During his voyages Zheng He made a map which records in detail his sea routes, including anchorages, reefs and shoals. None of these exploits would have been possible without the compass.

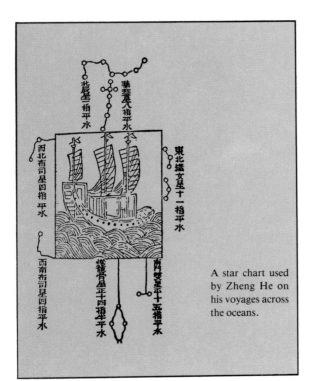

A star chart used by Zheng He on his voyages across the oceans.

Astronomy

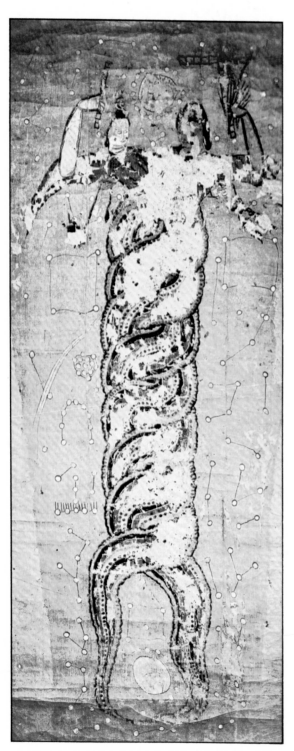

The earliest application of astronomy in China, in 2600 B.C. or thereabouts, is credited by legend to Huang Di, the Yellow Emperor, who established the 60 Year Cycle as the basis of the Chinese calendar. Such a calendar was adopted by the Xia, China's first dynasty (21st-16th centuries B.C.)

More than a thousand years later, around 1400 B.C. during the Shang dynasty which succeeded the Xia, the stars were used to mark the changing of the seasons. This is said in "The Canon of the Emperor Yao" in the *Book of History,* a history of China from earliest times said to have been edited by Confucius in the 5th century B.C. However, this must have gone on for a long time before, since the names of the "Fire Stars" (the constellation Scorpio) and "Bird Stars" (the constellation Hydra) were carved on China's earliest written records, the inscribed oracle bones of the 17th to 12th centuries. From the *Book of History* we learn that the spring equinox, summer solstice, autumn equinox and winter solstice were marked when the constellations Hydra, Scorpio, Aquarius and the Pleiades reached the meridian at dusk. The early royal courts had an "Official of the Fire Stars" whose job was to watch for their rising.

Many people believe that the old Chinese calendar was based solely on the movement of the moon and that the solar calendar was imported from the West. This is a misconception. The phases of the moon provided a convenient measure for the passage of time. But the oracle bones reveal that in China, even thirty centuries ago, the knowledge of the apparent movement of the sun was used as the basis for calculating the year.

Since the earth takes a little over 365 days to complete a course around the sun, and the moon takes about 29 1/2 days to move around the earth, it was a problem to fit them together. That they were combined very early into a solar-lunar calendar is shown by a mention, on oracle bones, of a

Silk painting found in a tomb dating to 500-640 A.D. in Turpan. Xinjiang. In the center are Fuxi and Nuwa, creators of the universe in Chinese legend, surrounded by constellations.

"thirteenth month." The *Book of History* also relates that the statutes of the Emperor Yao (c. 2500 B.C.) set a year of 366 days, with the four seasons regulated by an extra or intercalary month. The lunar months were 29 or 30 days.

Around 595 B.C. Chinese astronomers formulated the rule that an intercalary month inserted into 7 out of 19 lunar years made them equal to 19 solar years. This was 160 years before the Greek mathematician Meton discovered the same principle. The Greco-Roman world also used a combined solar-lunar calendar, but it was regularized only when Julius Caesar introduced the Julian calendar in 43 B.C.

China's records of certain heavenly phenomena are the oldest in the world, and the most detailed. Solar eclipses have been noted since the time of the oracle bones. The first to be definitely dated was that seen on February 22, 720 B.C., in the State of Lu, which is now southern Shandong province. As the records became more detailed, they served as the basis for predictions—which were carried on from about the 3rd century B.C. The greater ability to foretell eclipses reflected, in many ways, the general progress of Chinese astronomy.

Comets were called "broom stars." As in the Western world, it was believed that they foreboded disaster. Their passage was carefully noted, the earliest record dating from the autumn of 611 B.C. when a comet was seen in the constellation Ursa Major (the Great Bear). Between 204 B.C. and A.D. 1607, Chinese astronomers recorded every appearance of the well-known Halley's comet, which is seen about every 75 years.

China also has the oldest data on sunspots, observed as the sun was just rising. Her systematic records tell of 101 such phenomena occurring between 28 B.C. and the end of the 16th century. Europeans had seen them but did not realize that they were a part of the sun until the 17th century, after Galileo's observations through a telescope.

In studying the stars, the Chinese very early grouped the constellations along the ecliptic, or the plane of the orbit of the earth, into Twenty-eight

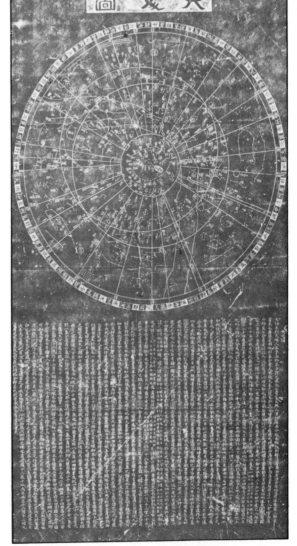

A star chart on stone in Suzhou.

In 1054 a Chinese astronomer observed and recorded the supernova explosion that gave birth to the Crab Nebula, a luminous cloud 13 light years across and 6,500 light years' distance from the earth.

Simplified armilla, a copy made in 1437. At the beginning of the Yuan dynasty, astronomer Guo Shoujing simplified the complicated ancient instrument by separating the circles of the horizontal coordinate system from those of the equatorial system. It reduces obscuration of the sky by the many rings of the old instrument.

Armillary sphere, a copy made in 1437. This ancient cast-bronze astronomical instrument which contains an observation tube, is a masterpiece of engraving and casting. It was used mainly for determining the equatorial coordinates and the longitudes and horizontal coordinates of celestial bodies.

Celestial Globe. This bronze astronomical instrument shows various coordinates of the celestial sphere, the apparent movements of celestial bodies and the positions of bright stars. Made in 1905 toward the end of the Qing dynasty, it is marked with 1449 stars.

On March 8, 1976, this meteorite fell in the suburbs of Jilin, a city in northeast China. Weighing 1770 kilograms, it is the largest yet found in the world. The scale is 30 centimeters long.

Mansions. These are so remarkably similar to the Twenty-eight Nakshatras of India that one must conclude that they had the same origin. Judging by the fact that many of the constellations were already mentioned in the *Book of Songs,* a collection of China's oldest songs and poems, the system of the Twenty-eight Mansions may have been known at the beginning of the Zhou dynasty (c. 11th century B.C.) and thus may have originated in China.

One of the great intellects of ancient China was Zhang Heng (A.D. 78 – 139), who was a contemporary of the famous Greco-Egyptian astronomer Ptolemy. Zhang Heng propounded the revolutionary theory that the universe is infinite in space and time. The earth, he said, was at the center, suspended like the yolk in an egg. Basing his observations on the position of the stars as seen from the Han dynasty capitol of Luoyang in Henan province, he took the Polar Star as one end of the axis on which the great sphere of the sky revolved over and about the earth. This was his explanation for the apparent movement of the sun about the earth.

Zhang Heng fought hard though unsuccessfully against the older Chinese idea that the earth was flat and the sky a dome. He invented the armillary sphere, a kind of celestial globe on which metal rings represented the paths of the sun and other important celestial bodies. Powered by a water-clock, a time-keeping device which had been developed much earlier in China, it could be called the world's first planetarium, and he used to delight visitors by showing them movements on the machine identical with those in the sky. One of the most ingenious of his many other inventions was a bronze device for detecting earthquakes. He was also the first in China to consider the moon more or less as we do today, as a non-luminous body reflecting the light

On a wooden chest unearthed from a tomb 2,100 years old are painted the Twenty-eight Mansions. The earliest record of the Twenty-eight Mansions so far discovered, it shows that the system had already been perfected at that early date.

of the sun. Eclipses of the moon, he explained, were caused by the shadow of the earth upon it.

The law of eclipses was finally formulated for China by Zhang Zixin in the 6th century. Fleeing from the wars and anarchy which characterized the struggle with former nomadic peoples for control of the northern part of the country during that time, he took refuge on an island where he spent thirty years studying the movement of the sun, the moon and five of the planets with the aid of an armillary sphere. He said that:

> ...the sun moves slower than average after the spring equinox but faster after the autumn equinox. The solar eclipse occurs when the new moon passes the sun within the limits of the solar eliptic. No solar eclipse will occur if it passes the sun outside of that limit. But a lunar eclipse may occur at every full moon if the latter happens to be near the node.

These discoveries were helpful in predicting eclipses.

During the Northern Song dynasty (A.D. 960 − 1127) the mariner's compass became known, first as an astronomical instrument and only later as an aid to navigation. Its form then was a disk to which was fixed a magnetic needle pointing, as Shen Kuo, a famous astronomer of that time, described in one of his books, "south deviating slightly toward the east." Also in the Northern Song period the waterclock was used in China to measure the hours.

During the Yuan dynasty (A.D. 1271 − 1368), at the time of the Mongol empire, China's astronomy benefited greatly from the contributions of the Arabs, transmitted through contacts with Central Asia and eastern Europe. In 1267 Emperor Kublai Khan was presented with Arabian instruments for studying the skies by Jama al Din, a Persian astronomer. Not long afterward, Kublai Khan also received instruments built by the leading Chinese astronomer Guo Shoujing (A.D. 1231-1316). They were an armillary sphere and a compendium instrument which was used to observe the positions of the planets. Guo produced more than 10 instruments and 14 books. Most of his heritage has been lost, but copies of the two instruments he gave the emperor still exist. As the chief astronomer as well as chief engineer for the Yuan dynasty, Guo Shoujing initiated the grand scheme for measuring the latitudes for major cities across the empire from Korea to Kunming in Yunnan province. He set up 22 observatories and calculated the latitudes of 27 cities, based on the height of the Polar Star.

Guo made the calendar more precise, calculating the solar year as 365.2425 days, only 24 seconds short of its true length. This enabled him to predict eclipses of the sun accurately to within an hour. He was the first man to formulate the theory of interpolation and utilized it to improve his calendar. In this respect he antedated Isaac Newton by 400 years.

Shen Kuo (1031-1095), a scientist who invented many astronomical instruments.

After Guo Shoujing's day, astronomical studies in China made little progress. Fearing threats to their power, the emperors of the Ming dynasty (1368 − 1644) closed China's land frontier's, and after Zheng He's famous sea voyages to India and Southeast Asia were cut short by jealous politicians inside the court, China was deprived of all contacts with the rest of the world, and science and the arts lost their originality.

The Twenty-Four Solar Terms

Chinese calendars today give dates not only of the internationally used calendar, but also the lunar calendar popular in the Chinese countryside. The latter gives the 24 solar terms, which serve as a kind of farmer's almanac of practical value in planning agricultural work.

When men discovered that the change of seasons had something to do with the sun, they began to apply their astronomical knowledge to determining the solar terms. As early as the Spring and Autumn Period (770-476 B.C.) the four solar terms of the summer and winter solstices and the spring and autumn equinoxes were recorded in books. By the time of the Western Han dynasty (206 B.C.-A.D. 24) the complete 24 solar terms as we have them today had been developed.

The 24 solar terms represent exactly the sun's position on the ecliptic. This apparent orbit of the sun is divided into 360 degrees. Every 15 degrees eastward marks a solar term. As the speed of the sun's movement along the ecliptic varies at different times, the number of days in the solar terms varies accordingly. For instance, the sun moves faster before and after the winter solstice, so each of the solar terms around this period is of a little more than 14 days; whereas at the time of the summer solstice, the sun moves more slowly, so each term is 15 plus days. A complete cycle of the solar terms equals exactly the time the earth takes to revolve once around the sun, or one year. The diagram shows the earth's position along its orbit at every solar term.

The names assigned to the 24 terms represent the climatic characteristics in China, especially in the Huanghe (Yellow) River basin. Here a monsoonal climate of four distinct seasons prevails, with cold winters and hot summers.

Four terms — "Spring Begins," "Summer Begins," "Autumn Begins" and "Winter Begins" — denote the commencement of the four seasons. Five indicate changes in temperature. Of the 12 months, January is the coldest and July the hottest. "Moderate Cold" and "Severe Cold" both occur in January, and "Moderate Heat" and "Great Heat" in July. After "Autumn Begins" comes "Heat Recedes," which shows that the hot days are over and the weather is beginning to turn cool.

Seven of the solar terms indicate changes in the atmosphere's moisture content during the year. "Spring Showers" signifies gradually increasing rainfall after spring sets in. "Grain Rain" denotes a marked increase in the amount of rainfall, meeting the need of the spring crops. "White Dew" means dewdrops during the night and so the weather is turning cooler. "Cold Dew" shows the weather will become even colder. "Frost Descends" and "Light Snow" mark the coming of frost and snow, signs around the beginning of winter, while "Heavy Snow" indicates winter has come in earnest.

The four remaining solar terms are named with reference to farming activities and how living things react to weather changes. "Insects Waken" is characterized by warmer weather, claps of thunder, and hibernating insects waking up. "Clear and Bright" announces the arrival of lovely mild weather when leaves and grass begin to sprout. "Grain Forms" tells us it is the time when the crops are growing strong and the wheat grains fill out. "Grain in Ear" ushers in wheat harvesting and summer sowing.

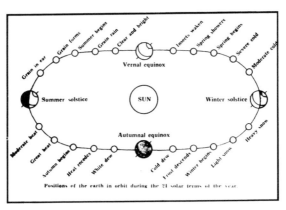

Positions of the earth in orbit during the 24 solar terms of the year

Chinese Medicine

How old is China's medical science? Its age can be gauged from one fact—"oracle bones" of the 13th century B.C. bear inscriptions describing various ailments of the human body. *The Book of Rites,* a manual of ceremonies written in the Zhou dynasty (11th c.-256 B.C.), records the court physicians' division of medical teaching into internal medicine, surgery, nutrition and veterinary practice.

The Yellow Emperor's Canon of Internal Medicine, which appeared during the Warring States period (475-221 B.C.), systematically presented what was known in China of physiology, pathology, diagnostics, treatment and preventive medicine. Theoretical bases for diagnosis and treatment were given. The earliest book in this field, the *Canon* contains a surprising amount of scientifically valid facts and views. Bian Que, a noted doctor at that time, was the first man in the

'The Yellow Emperor's Canon of Internal Medicine' is China's oldest extant treatise on medicine. Shown here is the chapter on acupuncture and moxibustion.

An exercise chart with over forty movements found in a Han dynasty tomb dating back 2,000 years. The Chinese people have long realized the value of physical exercise in curing and preventing disease.

Portrait of Li Shizhen (1518-1593), outstanding pharmacologist of the Ming dynasty who spent 27 years completing his Compendium of Materia Medica. The book was of great value to the future development of pharmacology. A doctor of Chinese medicine diagnoses a case.

world to use the pulse in diagnosis.

In the first century came Shen Nong's *Canon on Materia Medica,* China's earliest book on pharmacology compiled systematically. It listed 365 different kinds of herbal medicine divided into three grades, described their places of production, methods of collection and preparation, prescription, and medical use. It recorded, for example, that asthma can be treated with Chinese ephedra (*Ephedra sinica*) and malaria with the roots of antipyretic dichroa (*Dichroa febrifuga*). Both are still being used today.

Hua Tuo (?-208), a famous doctor in the 2nd century, applied an anesthetic powder in abdominal surgery. *Treatise on Febrile and Other Diseases* by his contemporary, Zhang Zhongjing, was the first medical work containing a fairly thorough description of theory and clinical experience.

Along with the publication of more medical works and the appearance of specialized fields in medicine, the Imperial Medical Academy — the earliest in the world — was set up in China in the 7th century under the Tang dynasty (618-907). It had over 300 students, studying in four departments: medicine, acupuncture, massage and "spells" (a practice introduced under Buddhist influence which did not play an important role. Chinese

medical science having thrown off the influence of sorcery almost a thousand years earlier). The department of medicine was subdivided into sections for medicine, surgery, pediatrics, moxibustion and the ears, eyes, mouth and teeth. The textbooks were authorized by the government and the period of study was from three to seven years.

The existence of hospitals in China dates back to A.D. 510 when one was set up for the sufferers from a serious epidemic in Shanxi province. By the end of the Tang dynasty, there were government-organized hospitals for the poor, as well as leper hospitals.

Also during the Tang dynasty a revised edition of the *Materia Medica* was produced, containing notes on 844 medicines. Further valuable work was done in the 11th century at the time of the Song dynasty (960-1279) by an official body known as the Bureau for the Revision of Medical Books. By this time the invention of printing had led to a fairly widespread distribution of medical literature.

Diagnosis in Chinese medicine is done by means of the "four examinations" — to observe (the complexion and fur on the tongue), smell (the odours given off by the patient when talking, breathing or coughing, and the smell of his excreta), ask (about the condition of the patient), and

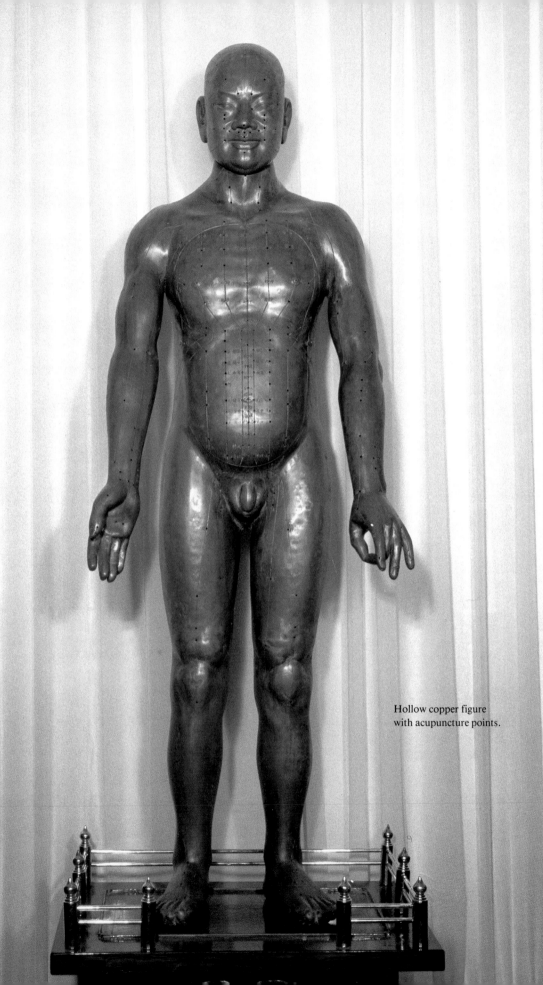

Hollow copper figure
with acupuncture points.

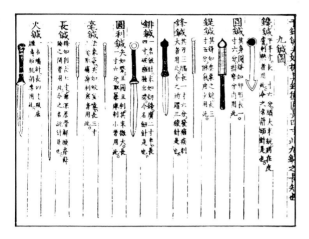

The earliest records of acupuncture were found on oracle-bone scripts of the 14-11th centuries B.C. Wang Weiyi, a noted doctor of the Song dynasty, compiled the 'Illustrated Manual on Acupuncture and Moxibustion Points on Bronze Figures' and supervised the casting of two life-size figures with acupoints marked.

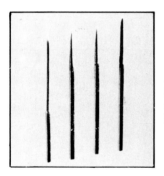

Acupuncture needles found in a 2nd-century Han dynasty tomb in Mancheng, Hebei province.

An American doctor learns acupuncture from his Chinese teacher.

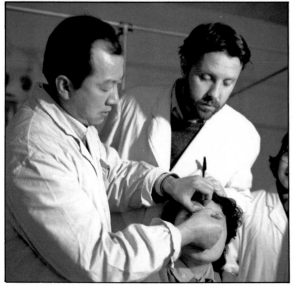

feel (the pulse).

Between the 10th and 14th centuries there was increasing contact between China and Arabia and eastern Europe. The divisions of medicine were increased to about thirteen including internal medicine, surgery, gynecology, orthopedics, acupuncture and moxibustion, neurology and infectious diseases. Acupuncture and moxibustion became very popular then. Life-size bronze human figures were cast with holes on them at the spot where needles were to be thrust in for practicing.

After the 15th century, during the Ming dynasty, innovation resulted in the formation of a system of treating communicable diseases. Outstanding landmarks in this period included the famous *Compendium of Materia Medica* by Li Shizhen (1518-1593). Wu Youxing's new theories in etiology, discoveries in anatomy by Wang Qingren (1768-1831), and the widening use of inoculation against smallpox which was invented in the mid-16th century. It consisted of extracting the contents of the pustules of a smallpox victim and either blowing it into the nostrils in powdered form or applying it to the nose with cotton.

Autopsies were done as long as 2,000 years ago. *The Yellow Emperor's Canon of Internal Medicine* recorded the length of human skeletons and blood vessels, and the position, shape, size and weights of internal organs. That the blood constantly circulates was stated in the *Canon,* which also dealt with the relationship of the movement of the aorta, breathing and the pulse rate.

Before Li Shizhen's time, there had already been quite a few pharmacological works in China. Among them are Shen Nong's pioneer *Canon on Materia Medica, Revised Materia Medica* issued in A.D. 659, and *Classified Materia Medica* completed at the end of the 11th century during the Song dynasty, which increased the collection of herbs to 1,746.

Li Shizhen completed his *Compendium of Materia Medica* in 1578. It was the world's most detailed classification up to that time. The labor took him 27 years of research, during which he traveled through more than half of China and made three revisions of the book. In the work, he reorganized the medicinal entries of previous books on pharmacology by reclassifying and regrouping many of them and cutting out repetitious or dubious ones. Some erroneous and ambiguous records were corrected and 374 new items added. His book lists 1,892 medicines illustrated by 1,000 drawings — 1,094 from botanical sources, 444 from zoological sources and 275 from mineral sources. It has been translated into Latin, Japanese, English, French, Russian and German and is known throughout the world.

Acupuncture and moxibustion make up an important part of Chinese medicine. Found among relics of China's New Stone Age (10,000 to 4,000 years ago) "pinning stones" were the predecessors of acupuncture needles. During the period between 770 and 221 B.C. the theory of *jingluo* (channels and collaterals) had been established, which was systematically illustrated in *Canon of Internal Medicine.*

Dispensers weigh out medicines.

As early as 500 B.C. Bian Que used acupuncture to save the life of the prince of the State of Guo, who was close to death from shock.

Acupuncture and moxibustion are different methods. Both are applied to points on the human body selected on the basis of the theory of *jingluo.* According to ancient Chinese medical theory, they cure diseases mainly by means of adjusting the relationship between *yang* and *yin,* promoting communication between channels and collaterals, regulating the vital energy (*qi*) and blood, promoting positive factors and eliminating negative ones.

Chinese medicine holds that the human body is a small world to which man must adapt himself. The theory of "agreement between man and nature" means that the human body must follow the law of nature — birth in the spring, growth in summer, harvest in autumn, storage in winter — so as to keep fit. Prevention has long been more stressed than treatment in Chinese medicine. One way to prevent diseases is to be active in physical exercises. Toward the end of the Han dynasty Hua Tuo worked out the Five Animal Exercises in imitation of the movements of tiger, deer, ape, bear and crane. These exercises helped cure and prevent diseases. Hua Tuo and his pupil Wu Pu lived as long as over 90 years. Later these bodily movements and deep breathing exercises were developed into *taijiquan* and *qigong* exercises that have played an effective role in improving people's health and preventing diseases.

Medicinal herbs.

Prescriptions of 2,000 Years Ago

In December 1972 a collection of 92 inscribed wooden slips was unearthed from a tomb in an earth cave at Wuwei county, Gansu province. Made of pine or poplar wood, they dealt with ancient Chinese medicine. Buried with them were a pottery urn and plate, a rod with a head in the shape of a dove, four other funeral objects and five coins of the *wu chu* type. Careful study and examination by archaeologists has shown that these finds date from an early period of the Eastern Han dynasty (A.D. 25-220).

Richly informative, the slips are apparently records of medical practice kept by physicians of that time. They comprise more than 30 prescriptions for treating various diseases in the fields of internal medicine, surgery, gynecology and ailments of the ears, nose, throat, eyes, mouth and teeth. The prescription usually begins by stating the disease, its symptoms, nature and causes. Then it specifies the medicines to be used, their dosage, the ways to take them and contraindications to their use. Finally, it tells how to prepare the drugs.

Classified clinically, in internal medicine the slips refer to fevers and diseases of the respiratory, digestive, circulatory, reproductive, urinary and

Tongrentang (the Universal Medicine Shop), located off Qianmen Street in Beijing, was founded in the mid-17th century by Yue Zunyu (1630-1688). Since then it has compiled a great many medical formulas, including the prescriptions of famous Chinese doctors once kept as family secrets. Its reputation rests on choice ingredients, careful processing, and high standards of purity.

nervous systems. In the surgical category, they mention ulcers and sores, carbuncles, dogbites and urinary stones. Also prescribed for are pharyngitis, deafness, nasal polyps, eye and dental diseases.

From the point of view of pharmaceutics, they mention about 100 different medicines including 11 made from animal, 61 from vegetable and 16 from mineral substances. Most are still used as ingredients in clinical medicine. Forms in which the medicines were administered include decoctions, pills, powders, plasters, drips and suppositories. The binding substances used for drugs in semisolid form included honey, milk, lard and cream. All forms incorporated features specific to Chinese traditional medicine. Pills coated with honey, a case in point, are still in common use in China today.

Some of the remedies were for internal and others for external use. The latter were to be smeared or plastered on locally, or poured into the nose or external ear.

The wooden slips also refer to moxibustion and acupuncture, giving the names and locations of several acupuncture points and "forbidden zones," and describing techniques of needle manipulation.

Chinese Ceramics

Ceramics is the general term for pottery and porcelain. In its long history of evolution in China, the technology progressed from pottery to porcelain, and from celadon and white porcelain to colored porcelain.

Pottery was commonly made and used in the neolithic period 7,000-8,000 years ago by people living in scattered villages on the banks of the Yellow River, in the lower and middle reaches of the Yangtze (Changjiang) River, and along the eastern sea coast. They probably discovered the technique when clay smeared on baskets as waterproofing was accidentally exposed to fire. This led to the deliberate shaping of clay pots which were first dried in the shade and then fired. Varieties included gray, black, colored, stamp-designed, thin-shelled and openwork. Throwing pottery on a wheel gradually replaced the earlier method of building it up by hand.

About 3,000 years ago during the Shang dynasty, it was discovered that coating the surface of the clay with silicate glaze prior to firing would give it an especially smooth and brilliant finish. Blue-green glazeware produced this way was the earliest form of celadon.

Between the 11th century B.C. and the 3rd century A.D. — a period comprising the Zhou, Qin and Han dynasties — white and colored glazed pottery became very popular. Glazes appeared in more colors: green, pale green, brownish yellow and chestnut brown. The final products were harder, finer and more translucent.

Pottery was no longer a material used only for making household utensils but a substance equally suited in its wide variety to the manufacture of bricks, tiles or works of art. Representative products of this period are the Qin period life-size pot-

Dog-shaped water pitcher *gui* (Neolithic period).

Yellow glazed vessel
(Northern Qi, 550-577).

Pottery kiln (left) and molds used to impress designs on pottery (right), of the early Shang dynasty (16-11th centuries B.C.).

Pottery figurines (Han dynasty 206B.C.-A.D.220).

Porcelain pot with elephant head spout and dragon handle (Sui dynasty 581-618).

Black pottery wine vessel *lei* (Neolithic period).

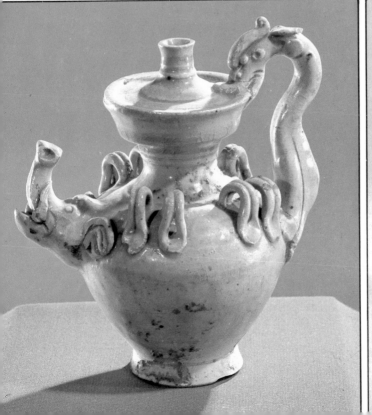

tery figures of horses and warriors, Qin bricks and Han tiles.

The period from the 3rd to 10th centuries A.D. (Three Kingdoms, Southern and Northern Dynasties, Jin, Sui, Tang dynasties) saw the maturation of Chinese ceramic technology. Shaping and glazing technologies rose to a much higher level. This is the period of the well-known "three-colored" (green, yellow, cream) Tang ware. also in this period the Yixing pottery in Jiangsu province began to make a name for itself. Yixing was later established as a ceramics center, achieving an international reputation which it still enjoys today.

Meanwhile, porcelain-making had also reached a high degree of perfection. The first true porcelain, derived from the greenish glazed celadon of the Shang and Zhou dynasties, appeared during the 3rd-5th centuries A.D., around the period of the Three Kingdoms and the Western and Eastern Jin dynasties. Later came white- and black-glazed porcelains.

From the 6th-10th centuries, during the Sui and Tang dynasties, the two kilns most representative of the factories of north and south China were the Ding potteries which manufactured white porcelain in the northern province of Hebei and the Yue kilns which produced celadon in the southern province of Zhejiang. The quality of their products was so fine, they were compared to snow and jade respectively. The Jingdezhen porcelain of Jiangxi province was of excellent quality. It was described as having "the blueness of the sky, the brightness of a mirror, the thinness of paper and the resonance of a chime-stone."

From the 10th-19th centuries porcelain was at its peak in moulding techniques, glazing materials, firing and quality and variety of painting and glazing. China had an unprecedented number of kilns and potters.

The greatest developments in porcelain during this period were in colors and glazes. Traditionally, patterns had been incised or imprinted on the body of a piece before glazing or firing. This was called underglaze decoration. The blue and white porcelain so famous throughout China and abroad was produced in this manner.

Based on the color glazing of previous dynasties, a new process was developed in the Qing dynasty (1644-1911). Red lead powder was mixed into the pigment and applied to the baked porcelain to create contrasts in color intensity, light and dark

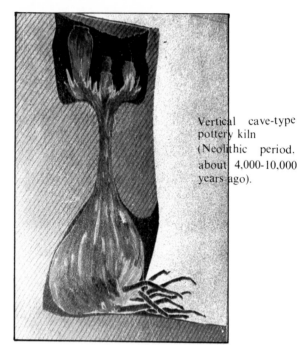

Vertical cave-type pottery kiln (Neolithic period. about 4,000-10,000 years ago).

Blue glazed porcelain bowl with golden flowers (Yuan dynasty 1271-1368).

Workshops at Jingdezhen, where porcelain was made for the former imperial court.

shades and a soft luster. The play of light and shadow on the surface of the porcelain produces a three-dimensional effect.

As early as the 7th century, Chinese ceramics were exported overland via the Old Silk Road or went by the southeasterly sea route to be sold in distant countries of Asia, Africa and Europe.

With advances in transport by sea, a single cargo of Chinese porcelain could number hundreds of thousands of pieces.

Colored pottery pot (Neolithic period).

Painted Designs on Ancient Pottery

Pottery used in the Neolithic period had designs imprinted mostly around the edge or mouth of the containers. This arrangement suited the lifestyle of primitive man, who placed his vessels on the ground, so that the designs were best viewed from above. Hence, these were distributed inside the rims of basins, on the shoulders of narrow-mouthed jars, etc. Applying this principle to the study of ancient pottery, one finds them decorated in a most attractive and unique manner. One small pot unearthed at Banpo when viewed at eye level seems to be adorned only with four rows of painted zigzags around its shoulders and belly. But when seen from above the ensemble forms an eight-petaled flower, with the mouth of the pot serving as the corolla and the lines on the lips of the pot as the outstretched stamens.

Warriors and horses
(Qin dynasty 221-206 B.C.).

Qin Pottery Army

In May 1974 while digging a well, the peasants of Xiyang village in Shaanxi province came upon the life-sized heads, hands and other parts of some terra-cotta sculptures. The village is located just east of the grave mound of Emperor Qin Shi Huang, the unifier of China (259-210 B.C.), in Lintong county 40 kilometers northeast of the city of Xi'an. A report to the county town soon brought investigators from the State Administrative Bureau of Museums and Archaeological Data and eventually led to an organized excavation of the site. A museum has now been set up there.

Preliminary excavation disclosed a buried army of life-sized warriors and horses made of terra cotta and charred remains of wooden war chariots. The warriors, holding real bows and arrows, cross-bows and spears, had once stood militantly in neat columns. Each chariot was drawn by four powerful horses.

The warriors stand 1.75 to 1.86 m. high, and the horses 1.23 m. They are in battle formation. In the forefront in the excavated section were three north-south rows of 72 warriors each; behind them troops were in 38 east-west rows with a rearguard and flank columns.

The characters "Gong Jiang" impressed on some of the figures are the same as those on some building materials found in the vicinity of the grave mound and known to be from it, another indication that they were made during the reign of Qin Shi Huang.

The 500-some warriors, carrying their weapons and wearing leggings and either armor or short gowns belted at the waist, exude a spirit of confidence and daring. Apparently based on real persons, the figures are well-proportioned and done in a realistic style with careful attention to detail. One can almost sense the flesh and bones beneath the cloth.

The figures have different expressions — fierce, bold, militant. There are warriors with broad foreheads denoting wisdom, warriors smiling with the joy of victory.

The horses are even more lively. The artist has created them with stylized forelocks and ears pointing forward, looking alertly into the distance as though if something appeared ahead they would neigh and gallop forward at a signal from their masters.

Illustrations showing porcelain kiln (left) and artisans painting designs on ceramics (right) from the book *Tiangong Kaiwu*.

Jingdezhen — the Home of Porcelain

Five-colored basin *xi* (Ming dynasty 1368-1644).

Porcelain jar with flower and dragon design and lid shaped like lotus leaf (Yuan dynasty 1271-1368).

Jingdezhen in Jiangxi province is synonymous with Chinese porcelain manufacture. Local craftsmen were producing pottery at the time of the Han dynasty 2,000 years ago, and porcelain kilns were built there in the 6th century A.D. during the Northern and Southern dynasties. By the 10th century during the Tang and Five Dynasties periods, improved firing techniques had made possible the manufacture of white porcelain so fine that it was known as "artificial jade." In the 11th century, one of the early Song emperors set up an imperial kiln at Jingdezhen to produce porcelains for use in his own household and in those of the nobility. As this happened during the Jing De reign period, the name of the porcelain city was changed from Changnanzhen to Jingdezhen.

Jingdezhen's products include both utilitarian and artistic porcelains. Most famous among the more than 3,000 varieties its kilns have turned out are the underglazed red and the blue and white of the Yuan dynasty, and the "Contesting Colors," "Famille Rose," "Cloisonne Colors," and "Five Colors" of the Ming and Qing dynasties.

A broad spectrum of glazes was developed at Jingdezhen. Commonly seen are Jun red, Ji red, Lang kiln red, rose red, green, yellow, blue, and black. Fired in a reducing flame, they are of very high quality.

Along with the development of its porcelain industry, Jingdezhen became a thriving commercial center in the Ming and Qing dynasties. To this day, the ancient center remains one of the most important pottery and porcelain making cities in China.

Mechanics and Engineering

"Dragon's backbone" water lift powered by a treadwheel.

Water mill used in ancient times.

IRRIGATION EQUIPMENT

One of the most ancient forms of Chinese irrigation equipment still in use today is the "dragon's backbone" water lift. Though less efficient than modern electrified pumps which are now replacing it, the device does save a great deal of labor. Its widespread use has contributed immeasurably to agricultural production throughout more than 1,700 years.

This water lift was one of the inventions of Ma Jun, an outstanding mechanical engineer during the 3rd century A.D. Implements which he invented and improvements he made on the silk loom and farm tools played an important part in expanding the productive forces of his time.

The "dragon's backbone" lift which he improved on the model of Bi Lan, a mechanical engineer in the 2nd century, was probably similar in structure to one described in the 13th-century *Book of Agriculture* by Wang Zhen, the earliest written account of such a device. It consisted of a trough with one end placed in the water. In it moved a series of flat pieces of wood, standing upright and closely fitting the trough. Through the center of each were attached pieces of wood connected by movable joints spacing the boards at equal distances in what resembled a continuous circular belt. This ran over a large wheel at the top of the trough and a smaller one at the bottom. When the former was turned by hand, the boards moved upward in the manner of the buckets of a modern ladder dredge, raising water from the river to the fields.

A variant was the "bucket wheel," a large, vertical waterwheel with bamboo or wooden buckets attached to its rim. The lower edge of the wheel dipped into the river and was turned by the current. The buckets were filled, carried upward and discharged into a flume running into an irrigation ditch. This device may also be driven by animal power. It came into use in the 7th century A.D.

MECHANICS

SHIPBUILDING

Historical records note that shipyards had appeared as early as the Spring and Autumn period (770-476 B.C.). Berths began to be used in the Qin dynasty, bringing shipbuilding to a new level. Yards combined berths with skidways for launching, a principle still used today.

A large shipbuilding site uncovered in 1974 in the center of the city of Guangzhou (Canton) on the southeast coast has been dated by archaeologists to the Qin dynasty (221-206 B.C.). Judging from the length and width of the two berths excavated, the first was used for ships with a cargo capacity of 30 tons and the second for those of 50 to 60 tons. A piece of wood unearthed from the site looks like a fragment of an oarlock base and has a tenon for inserting into a rib of the hull. The mortise-and-tenon was already being used to join planks and supporting members.

Three metal nails found indicate that planking was also nailed together. The size and arrangement of the shipyard are evidence that shipbuilding had thus advanced beyond the stage of fastening planks with ropes or leather thongs.

In the 16th century this windwheel was used extensively to lift sea water for salt-making.

Animal-drawn stone mill.

GRAIN PROCESSING MACHINES

One problem with rice is that it must first be hulled — the outer shell loosened by heating and removed — before it can have the bran polished off and be ground into flour. The water-powered tilt hammer introduced in the 1st century A.D. made hulling much less laborious. It was driven by a shaft rotated by a waterwheel. Cams fixed along the length of the shaft caused the hammers to rise and fall as it rotated. With the aid of this machine all the farmer had to do was to pour the grain into the mortar so it could be beaten by the falling stone pestle of the trip hammer. All the hard work was done by water power.

The flat millstone with the upper stone rotating against the stone beneath it was an early creation whose origin is lost in antiquity. Millet, rice, soybeans, wheat, barley, sorghum or corn were poured in through a hole in the upper stone and the flour which resulted spilled out through the gap between the two stones. The width of this gap determined whether the mill ground coarse or fine. These mills were driven by human, animal, wind or water power. In China, the latter was most readily available and water mills were built in which up to nine pairs of millstones might be driven through a geartrain by a single waterwheel.

巢车是我国古代军事
上使用的活动瞭望台，用
人力或畜力拖行，利用辘
轳将瞭望台绞起，人在台
上如同居巢中，所以叫巢
车。这种机械是春秋时期
劳动人民的重要发明。

"Nest cart" for military use during the Spring and Autumn period 2,500 years ago. The nest-like spotting platform could be raised with a windlass to give an observer a clear view of the battlefield.

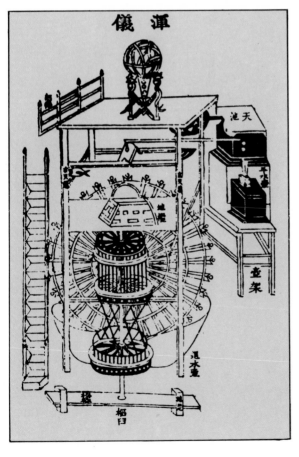

WHEELED TRANSPORT

The wheel was known in China at least as early as the Shang dynasty (16th-11th centuries B.C.) when kings went to battle in two-wheeled chariots. During the Warring States period (475-221 B.C.) the wheelbarrow was invented. During the early period of the Han dynasty (2nd and 1st centuries B.C.) four- and eight-wheeled carriages appeared. In the period of the Three Kingdoms which followed the Han in the 3rd century A.D., the southpointing chariot and odometer cart represented great technical advances.

The odometer cart was used for marking the distances traveled. The axle was fitted with a reduction gear train which drove a camshaft. Each revolution of the cam activated a wooden figure on top which struck a drum every *li* (half kilometer). A second figure struck the drum every 10 km.

THE EARLIEST SEISMOGRAPH

In 132 A.D. Zhang Heng, an outstanding scientist of the Eastern Han dynasty, invented the world's first earthquake-detector. It consisted of a closed bronze urn with a ring of eight dragons facing out from its rim. Between the jaws of each dragon was a bronze ball, and sitting around the base of the urn were eight bronze toads with their mouths gaping upward towards the dragons. Inside the urn was a rod or pendulum. When the urn tilted in the direction of any tremor that shook it, the pendulum pressed on the jaws of the dragon facing the tremor, causing it to open its jaws and drop its ball into the mouth of the toad below. Thus the clanging of the pendulum and the fall of the ball would indicate that an earthquake had taken place and also point the direction of its source.

Between 1088 and 1090 in the Northern Song dynasty astronomer Su Song and his associates designed and built a new three-tiered armillary 12 meters high and 7 meters wide. On top was an armillary sphere for observing the positions of the sun, the moon and stars. Below it was a celestial globe, turned by a mechanical system from east to west in line with the movement of celestial bodies. The entire system was powered by water from the tank at right. This armillary fell apart over the years. Fortunately Su Song described it in detail in his book *Essentials of the New Armillary*. A replica based on his description now stands in the Museum of Chinese History in Beijing.

China's first seismograph, invented by Zhang Heng in 132 A.D. When an earthquake occurs the tremor tilts the pendulum inside so that it presses on the jaws of the dragon facing the tremor and causes it to drop the ball in its mouth into that of the toad sitting below.

Other Direction-Finders

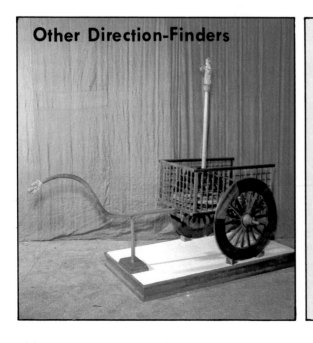

To know direction is a very ancient need. The semi-legendary Huang Di or Yellow Emperor, who dates from around 2 600 B.C., is said to have invented and drawn with his armies in battle a "south-pointing chariot" which, no matter where his course wound, would always indicate that direction. And old books say that such a chariot was used in the Warring States period (475 — 221 B.C.) but no one knew how it was made. To settle an argument that such a thing could exist, in the third century A.D. the mechanical engineer Ma Jun devised such a chariot operated by a gear mechanism. We are not certain how it was made, either, but the structure of one produced by the noted scientist Zu Chongzhi (429-500) was recorded in great detail. Such chariots were used until about the tenth century.

Ting cauldron for meat and cereals with human-face design (Shang dynasty).

Bronzeware

Bronze-making started in China as early as the Xia dynasty (21st-16th centuries B.C.). Contemporary sites of the Yan Shi Erlitou culture in Henan province have produced primitive knives and drills and, in rare instances, bronze vessels and bronze bells. This was the early stage of the bronze age.

Bronzeware of the Erligang culture in Zhengzhou, Henan province is representative of the bronze-making in the early Shang dynasty (16th-11th centuries B.C.). Carrying forward the

techniques of the Erlitou culture, it had made great advances in casting methods and variety. The shaping is accurate, the walls of the vessels thinner and more even. On the containers are animal and geometrical designs. Inscriptions are rare.

Bronze casting reached its peak in the later period of the Shang dynasty. Variety and quantity had increased. The shaping is more dignified and the designs highly intricate and refined. Frequently seen are the bird design, the cicada design, the silkworm design, and the cloud and thunder design. Cast on the containers are inscriptions related to ancestor worship and the conferring of awards.

The Western Zhou dynasty (11th century B.C.-771 B.C.) saw the gradual decline of slave society in China. Bronzes of the early Western Zhou dynasty differ little from those of the later Shang dynasty, but the middle and late periods produced such new types as the food vessels *Fu* and *Xu* and the ewer *Yi*. The pieces became lighter and less elaborate, and some are decorated with fairly simple ring and band patterns, double ring patterns or broad-figure bands. Wordy inscriptions were frequently used, however, some as long as three or four hundred characters. These usually concerned ancestor worship, calls for sending out troops, records of merit, the granting of official titles and conferring of awards.

The slave system was gradually replaced by feudalism during the Spring and Autumn period (770-476 B.C.). Less casting was done by the royal family of Zhou and more by the ducal states. New developments of this period were the techniques of inlaying gold, silver, copper and turquoise, and the birds, insects and writing on bronzes of the Changjiang (Yangtze) and Huai river basins.

The Warring States period (475-221 B.C.) witnessed the collapse of the slave system and the rise of feudal society. More utilitarian bronzes appeared—bronze mirrors, hooks, coins and seals. The bronze-making of this period is marked by improved gold and silver inlaying techniques and lively designs showing human activities, such as hunting and battle scenes and banquets.

Metalware, lacquerware and porcelain came into wider use during the Qin (221-206 B.C.) and Han (206 B.C.-A.D 24) dynasties, gradually replacing bronzeware. Nevertheless, craftsmen developed some new bronzes that were lighter, more refined and of greater practical value.

The human race entered the bronze age at different periods. In China, as recorded material and archaeological chronology show, the Xia, Shang, Western Zhou, Spring and Autumn, and Warring States periods all belong to the bronze age. Bronze is an alloy of copper and tin, characterized by a low melting point, supreme hardness and stable chemical properties. Chinese bronzes possess a fine and exquisite quality rarely equaled by other bronze-age civilizations. The detailed inscriptions on them are of considerable value for historical research.

Ancient Copper Mine at Tonglushan

An ancient copper mine was recently discovered at Tonglushan (Copper Green Hill) near Huangshi in southeast Hubei province, which is still an important copper mining center. The old mine stretches over an area of about two square kilometers. Excavations, which began in 1979, also disclosed nearby ancient smelting furnaces which had been covered by slag and so preserved over the centuries.

First estimates on the age of the mine indicated that the earliest pits were dug at the outset of the Spring and Autumn period (770-476 B.C.), with additional pits being opened late in the Warring States period (475-221 B.C.)

Carbon-14 tests confirmed the general estimate, but dated the first pits somewhat earlier than 770 B.C.

The ore pockets worked by the ancient miners contained an average of 5-20 percent pure copper.

The quality of ore samples and the amount of slag (some 400,000 tons)left around the mine place the quanitity of copper taken from the mine during its lifetime at roughly 40,000 tons.

The mine structure consists of vertical shafts of 40 to 50 meters deep, horizontal tunnels branching off them at the levels where ore veins are found, and additional vertical shafts which descend from many tunnels called "blind wells" because they are not open to the air. The remains of wooden water troughs reveal a relatively complete drainage system in the mine.

Tools and equipment left in the pits show the mining technology of this period. There were bronze chisels, hammers, chisels and hoes of iron, an iron chisel with four bamboo hoops fastened around its wooden handle, and fragments of rattan or bamboo baskets. Among the most important finds were two wooden axles, which were used as windlasses. By means of ropes attached to the windlasses, baskets

Gilded bronze figurine with a lamp from the Chang Xin (Eternal Fidelity) Palace (Han dynasty, 206 B.C A.D. 220).

of ore and waste materials could be brought to the surface and tools and supplies lowered easily to various levels of the mine.

Short bamboo slips burned at one end and found inside the tunnels were probably used as lights.

The furnaces recovered consist of a base laid half underground, a smelting chamber, and a top. Beneath each furnace is a ventilation passage. One furnace is elliptical in shape, some of the others are rectangular. There is an outlet for the molten copper and one of what was probably a pair of tuyeres. These furnaces could reach a temperature of 1200-1300°C.

Also found were stone hammers and balls probably used to crush ore, shallow holes filled with crushed ore of uniform size, pottery vessels, bronze adzes, copper ingots, slag, iron ore dust and kaolin.

Tonglushan is located on the shore of Lake Daye, which in turn connects with the Changjiang (Yangtze) River. A collection of copper ingots each weighing 1.5 kilograms (with copper content of 93% up) found at the bottom of the hill suggests that smelted copper may have been transported over considerable distances before being made into bronze objects.

The discoveries provide many insights into ancient production methods. It is now clear that during the Shang (16th-11th centuries B.C.) and Zhou (11th century B.C.-476 B.C.) dynasties mining, smelting and bronze-casting were each very specialized processes and showed considerable technical sophistication at a very early date.

The 'Lost-Wax' Technique

Discoveries of beautiful bronzes in the past few years testify to the artistry of ancient times. Examples are those unearthed at Marquis Yi's tomb in Suixian county, Hubei province in 1978. Among them are a 65-piece set of bronze chime bells, over a hundred sacrificial vessels and three thousand arrowheads. The total weight is approximately ten tons.

These bronzes were made by the "lost-wax" bronze-casting technique, developed during the Warring States period in the 5th century B.C. The principles were basically the same as present-day smelting and molding techniques. First a wax model was made as a perfect image of the desired bronze object. This was then placed in a box and covered with several layers of clay. When this had dried it was fired to harden the clay and melt out the original wax model, leaving a perfect mold. Bronze was poured into the space left by the wax model. When the bronze had cooled the clay mold was broken away, revealing a casting shaped exactly like the wax original.

The splendid bronze lions, elephants, cranes and fabulous beasts of the Palace Museum and Summer Palace in Beijing were cast by the lost wax process, a technique that is still used in modern industry.

Site of a copper mine of the Spring and Autumn period discovered by archaeologists at Tonglushan in Hubei province. A vertical shaft of the ancient mine (left).

Bronze galloping horse (Han dynasty).

An illustration from the book *Tiangong Kaiwu*, showing work processes in an ancient foundry.

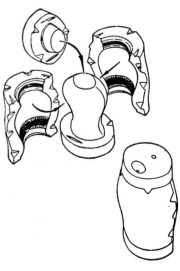

Casting process of the bronze goblet *zhi* of the Yin (later Shang) dynasty.

The Yong Le Bell

The Yong Le Bell—the second largest in the world—can be seen today at Da Zhong Si (Great Bell Temple) in Beijing's western suburbs. This bronze bell has a body 4.5 meters high and a maximum outer diameter of 3.3 meters. The loop for hanging alone is 1.1 meters high. Inscribed on its inner and outer surfaces are the Lotus Sutra and 16 other sutras totaling 227,000 characters and said to be in the handwriting of the Ming dynasty (1368-1644) calligrapher Shen Du.

This giant bell, long famed for its tone and exquisite workmanship, has been described in an ancient book as follows: "Struck day or night, it is heard several dozen *li* roundabout; its tone, harmonious and different from that of other bells, seems to come from both far and near."

The bell was cast during the Yong Le reign (1403-1424) of the Ming dynasty. Hence its name—the Yong Le Bell. It was first housed in the Wan Shou Si (Temple of Longevity), west of today's Purple Bamboo Park in Beijing's western suburbs. During the Qing dynasty (1644-1911) Emperor Yong Zheng (1678-1735) ordered it moved to where it is now.

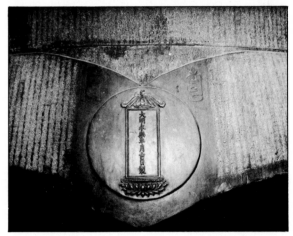

The moving of the bell, whose weight has now been estimated at 46.6 tons, was done quite ingeniously. In the winter water was poured on the road between the two temples and the bell was slid over the ice. The new bell tower was built up around it after it got there. The bell was placed on a mound of earth so that it stood at the height it would hang, and later the earth was removed. Qing dynasty emperors came here to strike the bell and pray for rain in years of serious drought.

The bell's thinnest part, right above the waist, measures 9.4 centimeters. Its sound bow or edge, 18.5 cm., is its thickest part, deliberately made so as to prevent cracking and improve tone quality. In fact, an outstanding characteristic of the bell is its musical tone. When struck it produces a harmonious chord, the main tones of which are very close to:

Chime-bells from the tomb of Marquis Yi (Warring State period).

Rare Find of Ancient Instruments

Displayed at the provincial museum in Wu-han, Hubei province are 7,000 tomb relics—musical instruments, bronzeware, weapons, articles of gold and jade, lacquerware and bamboo sheets, all from the tomb of Marquis Yi of the early part of the Warring States period (475-221 B.C.). But the most eye-catching among them is a set of 65 chime-bells, in many respects a masterpiece in history of Chinese music.

The chimes hang on an L-shaped wooden structure 7.48 by 3.35 meters long and nearly three meters high. Arranged in three rows, they include 19 small ones in the top row and 46 larger ones in the center and bottom rows. The largest weighs 203.6 kilograms and is 154.4 centimeters tall. This set of chimes is the biggest of its kind so far found in China.

The structure of wooden beams on which the chimes are suspended was so well constructed that through 2,000 years it did not give way under more than 2,500 kg. weight. The center and bottom rows are supported on the hands and heads of bronze warrior figures. The wooden beams are decorated with colored designs and have bronze end pieces with relief or openwork designs.

Gold inlaid inscriptions, mostly related to music, are found on every chime-bell. Tests have shown that if struck accurately on the correct inscribed position, the chimes will each produce two different tones. The center row of chimes—the main section—has a range of three and a half octaves. Those in the bottom row, with thicker walls, produce a deeper and more resonant sound and serve to play the accompaniment. The upper row of chimes was probably used to set the tune and to provide one or two supplementary tones during performances.

Although buried underground for 2,400 years, the musical properties of the chimes are still excellent, with fine tone color, broad tonal range and outstanding harmonic effect.

Music both Chinese and foreign, ancient and modern, has been played on them. A concert performance put on during the Spring Festival this year featured a replica of these chime-bells and was a great success.

Architecture

The Great Wall winds nearly 6,700 kilometers over deserts and mountain ranges from Gansu province to the shores of Bohai Bay.

The earliest vestiges of Chinese architecture, excavated at Hemudu village in Zhejiang province in 1973, date back to primitive society 7,000 years ago. Among the findings were wooden frames joined with mortise and tenon. This method was later to be used by ancient architects to erect imposing buildings out of thousands and even hundreds of thousands of pieces of timber.

The technique of rammed earth for building the foundations and walls of massive imperial palaces, gardens and tombs was developed as early as the 21st century B.C. By the third century B.C., buildings with brick and arch structures emerged.

The coming of Buddhism between the fifth and tenth centuries was followed by an upsurge in building temples, pagodas and stone caves with magnificent stone carvings, sculptures and murals. In the Song dynasty (960-1279), building styles became more elegant and refined, and new progress was made in the techniques of brick, stone and wooden frames. The last ancient architectural

boom was seen in the Ming (1368-1644) and Qing (1644-1911) dynasties when the styles of city layouts, imperial palaces, gardens and ordinary dwellings were finalized in the forms seen today.

The basic form of Chinese traditional architecture is the courtyard enclosed by buildings, still commonly seen in Beijing. This layout was fundamental to the dwellings of the people as well as the emperors. In the palace, the buildings were grander and more elegant, and many courtyards combined to form an intricate building complex. Typical of this is the "Forbidden City" in Beijing. The nearly 10,000 rooms in the Forbidden City are divided into a public front area and a private rear area. The emperors conducted important ceremonies and met their court officials in the front area. The rear served as domestic quarters for the emperors, empresses, concubines and eunuchs.

Man-made gardens, indispensable to Chinese architecture, first appeared in the 14th century B.C. Noted for refined craftsmanship, Chinese gardens are designed to blend with the natural

Design on the eaves of a traditional-style house.

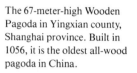

The 67-meter-high Wooden Pagoda in Yingxian county, Shanghai province. Built in 1056, it is the oldest all-wood pagoda in China.

The Zhaozhou Bridge built over the Jiaohe River between 605 and 617 in Zhaoxian county, Hebei province. It used to be a main artery for south-north traffic.

Qi Nian Hall, with three tiers of glazed tile roofs and a golden top, in the Temple of Heaven.

The ceremonial arch in front of the Lama Temple in Beijing.

scenery. Waterfalls, pools, shaped limestones, pavilions, terraces and covered corridors are the chief components of traditional gardens, which may not always be large, but provide a changing scene as one strolls along their walkways. The best gardens are found in Suzhou in East China. Court officials used to build elaborate gardens as retreats where they could indulge in drawing and painting and receive guests.

The layout of ancient cities was highly precise and regulated like a chessboard. In Changan (Xi'an) capital of the Sui and Tang dynasties, and Beijing of the Ming and Qing dynasties, the important government buildings were erected facing south on a north-south central axis. The south-facing imperial palace was located at the heart of the city. Main thoroughfares radiated away from the central axis. Workshops, factories, shops and markets, and dwellings of the people were grouped along the main streets to form smaller ones and lanes.

China's extraordinary rich legacy in bridge construction is perhaps best represented by her many stone arch bridges. The most famous of these is the Zhaozhou Bridge over the Jiaohe River near Beijing in Hebei province. Built in the Sui dynasty between A.D. 605 and 617, it consists of a single arch. The span of the arch, built up with limestone blocks, was 17.5 metres long, more than double the usual length — an extraordinary feat considering that neither steel nor construction machinery were available yet.

Generally the height of such a bridge is equal to the radius of the arch, but in this case, the arch is a segmental arc, making the bridge low and long with the crown of the arch rising only 7.23 meters higher than the abutments. Four smaller arches, two at each shoulder of the main arch, help to support the roadway and act as spillways. This design known as the open spandrelled arch became widely used in Europe only after the middle of the 19th century. The Pont-de-Ceret bridge in France was the first of this type known to be built outside China. It was erected in 1321 — 700 years later.

Among the magnificent projects undertaken by the ancient architects of China is the Great Wall, one of the wonders of the world.

Like a majestic dragon, the 5 to 10 meter high, 5 to 8 meter wide structure runs nearly 6,700 kilometers from Shanhaiguan (Between the Mountains and the Sea) Pass on the coast of Bohai Bay westward to Gansu province, passing through three

The Taihedian throne hall in the Forbidden City where Ming and Qing emperors conducted important ceremonies and received foreign ambassadors.

Humble Administrator's Garden in Suzhou, a city famous for its beautiful gardens constructed in traditional style.

A typical courtyard in Biejing.

other provinces (Shaanxi, Shanxi and Hebei) and the Inner Mongolian Autonomous Region.

Twenty states and dynasties contributed to building the wall. Parts of it were already being built in the seventh century B.C. The states of Yan and Zhao north of the Huanghe (Yellow) River, and others began to build walls in their own domains to prevent invasions by nomadic tribes from the north. After the Qin dynasty emperor Qin Shi Huang unified China in 221 B.C., he linked these sections and extended them to form the Great Wall. In the following millenium, many repairs and additions were made. The Northern Qi dyansty, whose capital was in Hebei province, in A.D. 555 conscripted 1.8 million laborers for reconstruction and repair of a 500-kilometer strip between today's Beijing and Datong in Shanxi province. The Ming dynasty (1368-1644) saw large scale reconstruction.

Excavations at the site of Zhou dynasty palaces dating back 3,000 years. With a total floor space of 1,469 square meters, this massive building complex was made up of many courtyards, similar to those popular in present-day northern China.

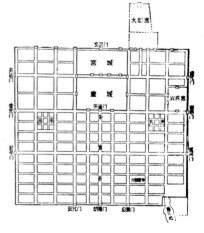

Drawing of the layout of Changan (today's Xi'an) capital of the Tang dynasty (618-907). Criss-crossed by 25 main thoroughfares, it measured 9,700 meters from east to west and some 8,600 meters north to south. The imperial palace was located in the northern part of the city.

Pavilions That Adorn Nature

The earliest pavilions in Chinese history were rest stops for postal and imperial couriers of the Qin and Han dynasties (221 B.C.-A.D. 220). Later similar structures were used to shelter stone tablets in front of imperial tombs. In palaces and temples, a building of the pavilion style sometimes covered a well or provided a frame for a bell.

Pavilions first began to be used to lend more beauty to gardens during the Sui dynasty (581-618). From the 14th century to the beginning of the 20th, during the Ming and Qing dynasties, pavilion-building developed greatly and the structures became increasingly intricate. The greatest variety in form and style is found in Beijing, Hangzhou and Suzhou where pavilions range from tiny ones to others covering as much as 100 square meters. More often seen are square, round and polygon designs, but some have unusual shapes — crosses, unfolded fans, double circles, triangles and half moons. Frequently pavilions are grouped in patterns.

The three main features of a pavilion are the foundation, the columns or pillars, and the roof. Built of brick or stone, the foundation is usually two or three steps above the ground. The pillars support an elaborate system of beams, cross-bars and girders all forming a design of beauty. A ceiling is seldom used, and this gives the interior an air of

Twin Pavilions in the Temple of Heaven.

Temple of Zhuge Liang in Chengdu, capital of Sichuan province.

spaciousness, allowing fuller appreciation of the artistry of the beam. A curved and peaked roof follows the outer lines of the pavilions. Larger structures have double and triple-layer roofs. The roof lines of most pavilions curve upward at the corner eaves. Most of the roofs are covered with tile, though sometimes they are of thatch to create a rustic effect.

In the south, complementing the clear waters and green hills, the pavilions are light and delicate. Most of them were built in the smaller private gardens of imperial officials. The coloring is subtle. Roofs are of bluish grey tiles, columns are deep brown. Built in rugged mountains, along the turbulent streams or in broad imperial gardens, north China's pavilions are generally magnificent in style, a bright spot in the scene. Their roofs are covered with glazed tiles in brilliant colors to brighten parks, imperial gardens and scenic places by their contrast with evergreen pines and cypresses of the dry north.

One of the four corner towers of the wall encircling the Forbidden City.

Pavilion of bronze and brass built in
1750 on Longevity Hill in the Summer Palace complex in Beijing.

Chinese Roof Styles

Gilded dragons decorate the top of a temple in the Qing dynasty Imperial Summer Resort at Chengde, Hebei province.

Of the centuries-old traditions of Chinese architecture, the curved tiled roof with upturned corners is the most outstanding. It lends piquancy and contrast to the plainer roof lines. In the *Book of Songs* (a collection of poetry from 770-476 B.C.), this style of roof is described as "bird's wings spread out ready for flight."

The style of sloping roofs varies with the climate, function of the building and the artistic inclinations of the designers. In the cold windy north, the most common two-sided roofs are thick and heavy, their eaves extending beyond the walls just enough to allow the winter sun to come in and give shade in the summer. In the warm rainy south, the roofs are thin and light. They have a steeper slope and a higher upturn at the corners of the eaves.

The most intricate and diversified styles of sloping roofs are found in the Forbidden City in Beijing. The dignity of Tiananmen, its south gate, is largely due to its huge double-tiered roof. The four corner towers of the wall surrounding the Forbidden City have a very rare kind of multi-sided roof. The roof of Taihedian, the imperial throne hall, is the largest of its kind — two eaves with four concave sides, topped with a long main ridge. The roof, covered with glazed tile, is nearly a meter thick and weighs about one ton per cubic meter.

The main building of the Temple of Heaven in the southern part of the capital is famous for its triple-eaved circular roof, covered with blue glazed tile and topped by a gilded oval-shaped ball. Without gables, this type of roof rises to a peak in the center.

Dwellings in the dry regions of China usually have flat roofs. Though not as varied in style as the sloping roof, they are built in many ways. Tibetans

Coffered ceiling of the Jiaotaidian Hall in the Forbidden City.

build two and three story houses in such a way that the flat wooden roofs of the lower floor serve as open platforms for the upper floors. In mountainous districts peasants use such roofs for drying grain.

On the loess plateaus in Shaanxi, Shanxi, Henan and Gansu provinces and Inner Mongolia, the people build the rooms of their houses with simple arched ceilings. The roofs are sometimes run together as a series of arches, but often the arches are covered over and the roof made flat.

Silk

Qing dynasty (1644-1911) drawing of weavers at work.

Sericulture—the production of raw silk and the raising of silkworms—is one of China's great contributions to the world. Legend has it that over 4,000 years ago, Lei Zu, wife of the Yellow Emperor Huang Di, was resting under the mulberry trees in the garden of her palace when she heard a rustling in the leaves. Looking up, she saw silkworms spinning their cocoons. She took one in her hand and found that the silken thread was shining, soft and flexible. "If we could wind it off and weave it into cloth," she thought, "how wonderful it would be!" And that, says the story, is how sericulture came to be invented.

This is legend. What is certain is that silk was used as weaving material more than 4,000 years ago (the weaving of hemp began some 2,000 years earlier). Evidence of this is a basket of woven silk excavated at a neolithic historical site in Zhejiang province in 1958. Jade carvings in the shape of silk-

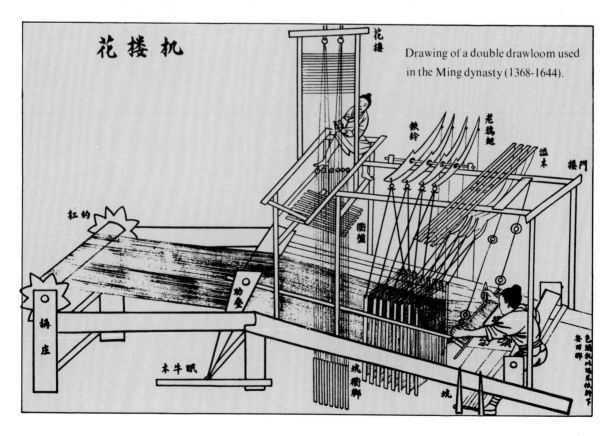

花楼机

Drawing of a double drawloom used in the Ming dynasty (1368-1644).

worms were funerary objects in the tombs of the slave-owning aristocrats of the Shang dynasty (16th-11th c. B.C.) and the characters for "silk" and "silkworm" have been found scratched on ancient oracle bones from that time.

Sericulture and silk weaving flourished over the centuries. Imperial government policy demanded that farmers in many places grow a certain acreage of mulberry trees to pay taxes in the form of silk. According to historical records, the Song dynasty Emperor Gao Zong (r. 1127-62) received levies and purchased as much as 1.17 million bolts of silk annually in Jiangsu and Zhejiang provinces alone.

Silk used to be the chief item of China's external trade. From the 1st century B.C. the product was carried over the Old Silk Road which started from the capital, Changan (today's Xi'an), ran through the Tarim Basin in Xinjiang and reached the Mediterranean, extending some 7,000 kilometers. China became known as "the country of silk." Merchants from Phoenicia, Carthage, Syria and other lands explored the sea routes to China in order to buy the precious fabrics. In the 2nd century A.D. a pound of raw silk fetched more than its weight in gold.

Via this route, the courts of the Han (206 B.C.-A.D. 220) and Tang (618-907) dynasties sent silks as presents to their vassal princes, to officials in the border areas, and to foreign monarchs or ambassadors. The silks were so prized by the recipients that they often left instructions to place these in their tombs after they died. A wealth of ancient Chinese silks have thus been found in tombs along the Silk Road.

Of the many varieties of silk fabrics — including satin, damask, muslin, gauze, velvet and tapestry — brocade woven in raised patterns with gold or silver thread was considered the most sophisticated. A number of excavations of brocade in ancient tombs during the last few decades has shown the advance of the brocade weaving techniques since it was first made in the 11th century B.C. The invention of the double drawloom spurred the production of brocade, making even more gorgeous patterns possible.

The four most renowned kinds of brocade to this day are Song brocade representing the style of the Song dynasty; Shu brocade of Sichuan with a history of over 1,000 years; yun (cloud) brocade of Nanjing in Jiangsu, characterized by elegant and colorful patterns; and elaborate Zhuang brocade made by weavers in the Guangxi Zhuang Autonomous Region who use cotton or linen yarn as warp and silk as weft.

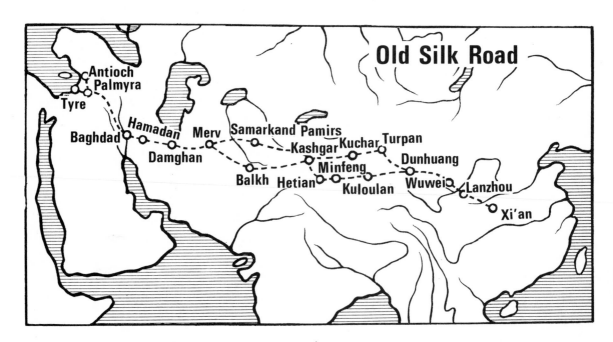

Shoes of colored silk dating from the 4th-5th centuries.

Design of highly stylized elephant on brocade woven in the Tang dynasty.

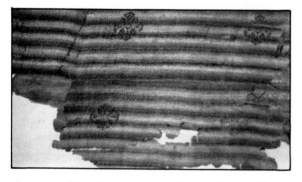

Fragment of Tang dynasty patterned brocade found in Turpan, Xinjiang, along the Old Silk Road.

Flower-and-bird design on brocade from the Tang dynasty (618-907).

Treasure-Trove of 2,200-Year-Old Silks

Silk quilt embroidered with dragon-and-phoenix design.

A fabulous collection of silks recently recovered from a tomb in Hubei province dating from the middle Warring States period (475-221 B.C.) includes representations of stylized animals and dancers, and also motifs such as phoenixes, hexagons, S-shaped figures and lozenges. Previously the earliest finds of this type dated from the East Han dynasty (A.D. 25-220).

The tomb was discovered by brickyard workers digging clay in Jiangling county in January 1982 just north of the Changjiang (Yangtze) River in south-central Hubei province. Now known as Mazhuan No. 1 (a contraction of the factory's name), the tomb is one of many from this period in the vicinity.

Eight kilometers to the southeast once lay a city named Jinan, which from 689-278 B.C. was the capital of the large and powerful state of Chu. So valued was the art of silk weaving in Chu that once, when it invaded the State of Lu which was known for its silks, one of its demands before withdrawing was that a hundred weavers, women silk workers and other artisans be handed over to Chu.

Though not large, the tomb contains a gorgeous array, including brocades in two and three colors, gauze, and pieces with fine embroidery. Several kinds of fabrics were often used in the same garment. Some of the plain weaves contain as many as 150 warp threads per centimeter, testifying to the high level of silk production at that time.

The brocades are of warp-pattern compound weave, in which warp thread passes over several weft threads to create the design.

One piece of brocade has a pattern unit 50 cm. long and 5.5 cm. wide, with several different motifs. They include mythical animals like the dragon, phoenix and unicorn, and dancers with flowing sleeves. Another has geometric shapes combined with the phoenix motif.

Even more valued than brocade in the Warring States period was embroidery for the greater labor expended on it and the more individual artistry. Fanciful animal designs are embroidered in shades of yellow, green, deep red, purple and dark brown on backgrounds of different colors. Phoenixes predominate, dancing, hovering with spread wings, standing full face or in profile, all vividly executed.

There are several garments of gauze "thin as a dragonfly's wing," as the saying went. On one beautiful piece is a large embroidered dragon-and-phoenix design combined with tigers in vermilion and black.

The Mazhuan No. 1 is the first tomb ever discovered of the Warring States period in which the body was prepared for burial according to the ceremonies and rites recorded in a book of that period. The coffin was filled with silks. There was a silk-covered coat with silk padding, then a quilt with an embroidered dragon-and-phoenix design. In a bundle tied with nine brocade ribbons were a dozen items, including brocade and embroidered quilts, gowns of silk padding and of gauze. The skeleton was clothed in three garments and laid over it were a piece of brocade and a silk skirt. Over the face was a gauze veil. The feet were tied together and in each of the hands, which were tied at the side of the body, was a roll of silk.

The most famous Chinese embroidery styles originated in Suzhou, Hunan, Guangdong and Sichuan. Suzhou embroidery attests to the wealth of opportunity, environment, materials and craftsmen to be found in the beautiful city set beside Taihu Lake and surrounded by green hills. Carrying on the traditions of the embroidered pictures of the Song dynasty of 1,000 years ago. Suzhou embroiderers execute "brush strokes" with fine thread and dense, even stitches.

The Hunan style also combines the skills of painting and needlework. Different kinds of stitches are used to imitate different brush-styles in gold thread and more than 700 colors. An example is the huge screen "Adding Flowers to the Brocade" featuring a peacock so lifelike it seems ready to step out of the screen.

Guangdong embroidery is noted for its wide

Embroidery

Chinese embroidery is no less venerable than silk-making. Archaeological evidence for embroidery dates back to the Western Zhou period (11th-8th centuries B.C.). A tomb excavated in 1974 in Baoji, Shaanxi province contained impressions of plaited stitch embroidery. By the Han dynasty (206 B.C.-A.D. 220) embroidery was widely used for decorating garments and articles of daily use. In the Song dynasty (960-1279) it developed into an art form specializing in the imitation of paintings and the calligraphy of famous artists.

Chinese embroidery began as a people's art, rich in symbols of the beauty of life and love of life. Traditional artifacts were pillows and perfume pouches embroidered with a box, the hand of Buddha or plum blossoms — supposed to represent lasting peace, numerous children and infinite happiness.

In imperial times the best folk embroiderers were often sent to work for the court. An example of their work is a jacket made for a Ming empress. On it are dragons, the character *shou* (longevity) and 100 boys catching birds, playing hide-and-seek, flying kites and wearing black gauze caps to impersonate officials. Embroidered in gold thread around the edges of the jacket are pine trees, bamboo, plum blossoms, rocks, and all kinds of flowers and plants.

Tiger 长白山虎

range of subject matter, bright colors and rich designs. "Waiting for the Moonrise at the Western Chamber" is a scene from the play *Romance of the Western Chamber*. In this embroidered picture Guo Yingying is shown standing by the garden gate, quietly waiting for her lover Zhang Sheng. Her eyes, measuring less than half a centimeter, are embroidered with dozens of threads of different colors which give them lifelike brilliance.

Sichuan embroidery, made in and around Chengdu, capital of Sichuan province, is known for its neat and tight stitches done with shiny and soft threads.

Research institutes in China's embroidery centers, especially those south of the Yangtze River, contribute a great deal to the development of the art of embroidery. Today's embroiderers use 50 different stitches and thousands of ways of combining threads of different colors. Reproducing the brush-strokes of Chinese painting, the artists in silk can depict landscapes, personages, flowers and birds with strength or extreme delicacy.

"Dragon robe" of Qing Emperor Qianlong (1736-1796).

Double-sided embroidery is made with a new technique. On a silk base as thin as a cicada's wing, the embroiderer employs thread only one forty-eighth as thick as that normally used as well as a variety of stitches to reproduce objects as fine and delicate as the tail of a goldfish. Thousands of knots and thread ends are skillfully concealed so that the finished work can be viewed from either side.

Traditional Chinese Handicrafts

The ingenuity of Chinese folk artisans has over the centuries found expression in countless varieties of traditional handicrafts, many of which still form an indispensable part of the daily lives of the Chinese people, and enjoy increasing popularity at international exhibitions and in world markets.

Bamboo Weavings

A fifth of the world's bamboo grows in China—300 varieties in a total area of 20,000 square kilometers. No one knows how long China has used bamboo. About 6,000 years ago, the Chinese character 竹 (*zhu,* or bamboo) was carved on the pottery of the neolithic Yangshao culture. In Zhejiang province 4,000 years ago there were bamboo baskets.

Bamboo is used most extensively south of the Changjiang (Yangtze) River. Almost every house has things made of it—beds, chairs, cases, baskets, brooms, chopsticks and even baby carts. Exquisite articles intricately woven of bamboo strips make inexpensive arts and crafts items. Peasants cut the tough outer layer of the bamboo into flexible and tensile thin strips and cleverly weave them into baskets, vases, bags, plates and lifelike animals.

People in the bamboo-growing areas often sit on summer evenings, chatting and weaving bamboo strips. In about five minutes a frog is made that can be made to hop just like a real one. Bamboo strips are also painted in red, black and blue colors, producing fancier patterns and more striking artistic effects. Apart from bamboo, palm fibre, cattail stems, wheat straw, rattan and twigs are also used in weaving.

Decorated Fans

A Han (206 B.C.-A.D. 220) carved brick shows a salt worker using a crudely-shaped fan to keep the stove burning. Later fans were used as dusters or as something for driving away flies or mosquitoes. Over the years fans have been made more and more beautiful. Ivory, ebony, sandalwood, mother-of-pearl have all been used as raw materials. The blades and end-pieces of a folding fan and the handle of a screen fan are often elaborately carved, gilded or inlaid.

Since the Ming dynasty (1368-1644) the folding fan has largely supplanted the screen fan as paintings and calligraphic work by famous artists made the fan a more valuable artistic work. Wang Xizhi (4th century), the most famous calligrapher in Chinese history, once took some bamboo fans from an old woman who was selling them and on each he wrote five characters. She was dismayed, thinking the stranger had soiled her goods. Smil-

Basket, jar and vase woven of bamboo.

Fans of different styles and materials.

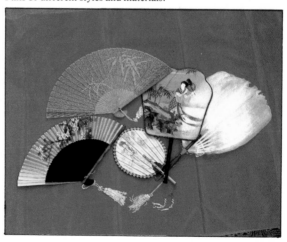

Painted clay dolls from Wuxi.

Cloth toy tigers from Shandong province.

these toys, and some of them have features that have come down unchanged for many hundreds of years.

The use of local materials—wood, clay, straw, bamboo and so forth—give the hand-made toys of different areas their special character and make them cheap to buy. Those made south of the Changjiang (Yangtze) River, which are mostly of reeds, grass or bamboo, have a greater delicacy and grace than the bolder and sturdier ones of clay and wood from the north.

The jointed dragon comes from the south. It is made of bamboo, cut into sections and strung together with wires. Its body has a carved pattern on it, painted vermilion and sky-blue. When the stick under its body is turned, the dragon jerks and twists in a highly lifelike way.

Creating intricate papercuts with scissors.

ing, Wang Xizhi said, "Tell your customers that the characters were written by Wang Xizhi." The fans were sold in no time. This story illustrates very well the increasing function the artist's decoration played in fanmaking.

Almost all great artists have left their works on fans, forming a treasure separate from their scrolls. Today, fans are being produced in many parts of China, each with a distinctive style. They may also be made of thin palm or cattail leaves, bamboo or wood strips.

Traditional Toys

Although Chinese towns and cities have many toyshops selling up-to-date things like trucks, airplanes and modishly-dressed dolls, the most common playthings are the handmade toys, of which there is a variety in every part of the country. They are generally sold in the streets by peddlers, or on stalls and booths at fairs and festivals. There is an immense amount of local custom and tradition in

Peacock—a multicolor papercut.

The little wooden figures from Zhejiang province on China's eastern seaboard are made from boxwood famous in those parts, a smooth-grained golden-hued wood, soft to carve. Children love these tiny representations of things familiar in their everyday life—the oxcart, the buffalo, the fishermen, boatmen, peasants winnowing grain or sowing. On the other hand, the fat, gaily-colored clay dolls from Wuxi are delightful representations of happy, exuberant childhood; they originate from folklore and are supposed to bring happiness. The Beijing "stick doll" is carved from a single piece of wood. It has a round comical face and its head is movable.

Papercuts

Colorful paper cut-outs pasted on windows, doors and walls are a traditional decoration in Chinese rural homes to mark festivals, weddings, birthdays and other joyful events. Transparent against the sun by day or silhouetted against the white window-paper by lamplight at night, they add gaiety and charm to the festival scene.

Papercuts dating from between 514 and 551 A.D., in the period of China's Northern and Southern dynasties, were among the objects found by archaeologists at Gaochang in the Xinjiang Uygur Autonomous Region in 1959. The design of facing pairs of horses, rendered with great skill, indicates that the art had already reached a high stage of development during the 6th century. Through the centuries they have generally been the work of peasant women and a professional trade as well.

The traditional designs of Chinese papercuts were largely symbolic. A pair of mandarin ducks swimming side by side denotes married love. A crane suggests longevity. Homophones of Chinese characters of different subjects are borrowed to animate designs. A magpie conveys the feeling of joy (the character for "magpie" having the same sound as the one for "joy"). New designs are linked with present-day life of the people. The desire for modernization, for instance, may be shown through a tractor in the field driven by a peasant woman.

There are many kinds of papercuts: in paper of one color, or painted, or combining papers of different colors pasted together, and also in very thin copper-foil. Styles vary regionally. Designs from the south tend to employ fine, delicate lines while those from the north feature broad, forceful ones. Some types are done with scissors and others

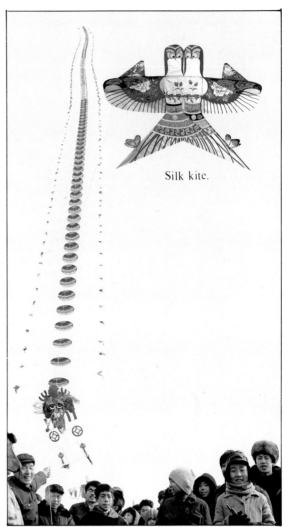

Silk kite.

Flying a dragon kite with a long tail.

with a two-edged knife.

Today's papercuts are used for wall decorations, bookmarks, greeting cards, postage stamps, calendars and headings for newspapers and magazines. Many illustrations and cover designs for books are based on papercuts. So is a totally new type of animated film, unique to China, which has won wide acclaim.

Kites

In very ancient times the Chinese name for kite was *zhi yuan* or paper bird. Then, in the period of the Five Dynasties (907-960) a man called Li Ye is said to have flown a kite with a small bow of bamboo attached to it, strung with a silk ribbon. When the wind vibrated the ribbon, the kite emit-

Cloisonne food vessel.

Enjoying the Lantern Festival, which falls on the 15th day of the first lunar month (February).

ted a musical sound and since that time kites have been called *feng zheng* or aeolian harps. Throughout the centuries, they have been put to various uses, not all of them recreational. In the sixth century they were flown for military purposes, to send signals from a besieged city. In some places they were thought to have magical powers of healing and were flown to cure patients of fever.

But it is as a pastime both for grownups and children that kite flying has remained popular. China has kites made in many shapes—figures, fairies, the Monkey King, flowers, birds, fish, worms, butterflies, goldfish, eagles, swallows, bats, dragonflies, centipedes and many others.

The two types in China are the "soft wing" and "hard wing," so named after the method of making them. The "soft wing" type of a smaller size can be taken apart and put into a box when not in use. The "hard wing" type is made with body, head, and wings all in one structure. Its frame is covered with stouter paper or sometimes silk. It can be flown in a stronger wind and stay in the air for a long time.

Festival Lanterns

Even now when electricity is widely used and candles are disappearing, people love to hang out lanterns for decoration at festivals and other times of rejoicing. Historical records show that big lantern festivals were held as early as the Tang dynasty (618-907). They tell of the crowds who thronged the capital for three nights to watch a giant wheel, decorated with silk and hung with fifty thousand lanterns, beneath which young women stood singing as they made it revolve. When the imperial capital moved to the south in the Southern Song period (1127-1279), Suzhou, Hangzhou and Fuzhou became centers of lantern making, and have remained so ever since.

Palace lanterns are often elaborate and costly, with carved and gilded frames covered with silk, gauze, glass or horn and hung with jade pendants and silk tassels. Many such lanterns hundreds of years old still hang in their original places in the Palace Museum in Beijing and elsewhere. They are widely used in modern buildings for interior and exterior decoration as well.

As for the people's lanterns, their variety is endless. There are fish, flowers, birds, insects, boats, animals, butterflies—hundreds of shapes, often made of bamboo and covered with oil paper or gauze. Some are carved out of real fruits or vegetables. Tiny ones are made from egg-shells, their four little doors propped up with minute green and gold posts and beams. Some lanterns are made from blocks of ice.

The Art of Carving and Sculpture

Chinese carving and sculpture use such materials as jade, wood, lacquerware, ivory and stones.

HANDICRAFTS

Jade carving is a general term which includes carving of many kinds of semi-precious hard stones—turquoise, coral and the green or white jadeite. The beauty of the stone and its indestructibility have made jade valued in China as the symbol for virtue and immortalized it as such in classical literature. The art of jade carving reached great heights at different periods of its 3,000 year history. In the first millenium B.C., one famous carving was considered worth 15 cities by the King of Zhao.

A painter can create whatever form he wishes on paper. But the work of the artist in jade is determined by the nature of his stone. The jade carver's clever hands turn the stones into a myriad of forms—landscapes, vessels, flowers, birds, animal and human figures. In the past few decades, the jade craftsmen have been able to create new designs and ways to use the material for unique new masterpieces.

Stone carving started in prehistoric times when early men made stone tools. The most famous and the finest works in stone carving are found in caves at Yungang in Shanxi province and Longmen in Henan province. Together, the two places provide a consecutive record of the development of Chinese sculpture from the time of the Northern Wei dynasty (386-534) down through the Tang dynasty (618-907).

The oval-shaped chambers of Yungang, built during the latter half of the fifth century, contain both reliefs and statues of seated or standing figures, the largest of which is 17 meters high. A wide variety of postures and expressions are executed in sweeping vigorous lines. The center of attraction is a seated Buddha. Its full, dignified face, broad shoulders, elongated ears, straight nose and rather thin lips smiling faintly produce an overall effect of austere strength as well as repose and contemplation. Cut out of the limestone hills on either side of the scenic Yi River, the 2,000 grottoes of Longmen contain some 14,000 images, the smallest two centimeters in height, the tallest, 17 meters. Eighty percent of them are Tang images, characterized by animation and vivacity and reflecting the spirit of that age. Chiselled out of the face of a cliff, the principal Buddha is 17.14 meters high, with a head four meters long and ears 1.9 meters long. The face, with crescent-shaped eyebrows and a gentle smile, has a warm, very human expression. Sculp-

tures of devotees create a mood of deep reverence for the Buddha. Heavenly kings and warriors are shaped in an exaggerated way, some with large heads and short legs. High cheekbones and protruding eyes heighten their powerful masculinity.

Miniature ivory engraving takes one into an entirely different dimension. One needs to look through a high-power magnifying glass to see a perfect replica of some treasured scrolls of Chinese writing or a painting engraved on a piece of ivory 5 millimeters in diameter—the size of a grain of rice. Microengraving has been done in China at least since the Western Zhou dynasty (11th century-771 B.C.). By the period of the Northern and Southern dynasties (5th-6th centuries A.D.) it was very pop-

An artisan paints a large lacquerware vase—Fuzhou, Fujian province.

ular. In recent years, some Chinese artists have succeeded in practicing their microengraving art on human hair. Shen Weizhong, a young ivory-engraver from Suzhou Arts and Crafts Research Institute engraved the words "We have friends all over the world" both in Chinese and English on two strips of hair, each 4 millimeters in length, and stuck them to an ivory globe model the size of the head of a match. He has engraved an entire poem by the famous Tang poet Zhang Ji on a 5-millimeter-long strand of hair.

CHINA: 7000 YEARS OF DISCOVERY

China's Ancient Technology

SPECIAL SUPPLEMENT

Profiles of Folk Craft Exhibition Demonstrations

Batik Cloth Prints

Wang A'yong

W ANG A'YONG, 36, is a Miao nationality woman from a mountain village in Danzhai county, Guizhou province. Her graceful and beautiful folk art has been practiced in southern China since the Han dynasty (206 B.C.-220 A.D.)

In batik peasant prints, patterns are traced with molten wax on white cloth. When the cloth is dyed, the wax keeps the dye away from the covered parts. Afterward the wax is melted away with boiling water, leaving the white patterns to stand out distinctly against the dyed background color (traditionally dark blue, but now one or more other colors are often used).

Miao women make beautiful dressings for weddings and other festive occasions with print materials of their own design. Skill and originality in batik dyeing is highly prized — and a quality that men still look for in a bride.

Wang A'yong learned her craft at an early age from her mother, and is now considered one of the best contemporary practitioners. Her design themes are inspired by the natural charm of her native place. Flowers, birds and fish in bold and fresh styles are common motifs. She is deft at tracing patterns freehand without making initial drafts on paper. She is now passing on her experience to her own children and other young people.

Embroidery

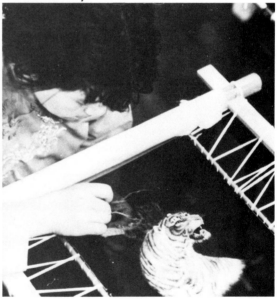

Peng Jianchun

P ENG JIANCHUN, 40, is a noted embroidery artist of the Hunan embroidery Research Institute in Changsha, Hunan province. Generations of her family have been embroiderers, and as a young girl Peng pestered her mother for lessons. To make her point, she secretly tried out a few stitches on her mother's embroidery frame, for which she was roundly scolded. But Peng kept practicing on her own, and soon was bold enough to add a whole leaf to her mother's work. Convinced at last of her daughter's sincere interest, the older woman agreed to teach her systematically. Peng's career got another boost when, at 16, her work came to the attention of a veteran embroidery artist.

Peng's regular "palette" of silk threads includes thousands of different shades, and each piece may employ several hundred. Her favorite motifs are human figures, lions, tigers, flowers and birds. She is an expert at the extremely difficult double-sided embroidery (which results in a complete perfect picture on each side of the fabric, with no ends or knots showing). A 1979 piece with a pug-dog on one side and a kitten on the other, done on a length of transparent nylon as thin as a cicada's wing, was recognized as a master work of Chinese embroidery. Visiting Hongkong that same year as a member of the Chinese garment and embroidery delegaton, her demonstrations of her skill drew widespread admiration.

In her spare time, Peng loves to attend Hunan folk drum opera performances. But many of her "spare-time" activities are actually related to her work. The trips to zoos to watch the birds and animals, the flowers and goldfish she raises, and her visits to painting exhibits all give her ideas that ultimately find their way into her embroidery. Her husband, a news reporter, does much of the housework so that she can spend more time on her art.

Dough Modeling

LANG ZHILI is only 41, but for almost three decades her nimble hands have been creating lifelike figurines, beautiful flowers and natural scenes out of dough. The special dough used in this folk art of northern China is made of wheat and glutinous rice flour mixed with honey, preservative chemicals and dyestuffs of various colors.

Lang's specialty is historical or fictional personages set against backgrounds of birds, flowers and animals. Her human figures are vivid and lifelike in expression and posture, with light, subdued make-up and matching attire. Her compositions are simple, but harmonious and balanced—the result of careful planning.

One of her most popular and critically acclaimed series is of characters from the classic

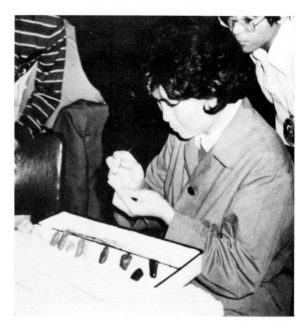

Lang Zhili

novel *A Dream of Red Mansions.* For a scientific documentary film, she made more than a hundred mushrooms of different kinds, which were much admired by botanists for their realism. Her fine miniatures, the details of which can only really be appreciated through a magnifying glass, are admired both in China and abroad.

Lang Zhili's first lessons were from her father, himself a noted Beijing dough modeler. At fourteen she became a student at the Beijing Applied Arts Research Institute, and some of her early works were displayed at an international young people's art exhibition in Moscow. Her father is now one of her biggest fans: "She's quite surpassed me—she's got really original ideas." A busy artist and designer with the Beijing Applied Arts Plant, she still finds time to coach youngsters in her craft at the city's Children's Palace. Many of her works have been shown abroad, and the Chicago trip will be her second to the U.S. The first was a demonstration visit to Hawaii in 1981.

Porcelain Sculpture

KANG JIAZHONG, now 50, was born in Yibin county, Sichuan province, and showed a liking for art from the time he was quite small. He entered the Chongqing Art College at 15, and de-

Kang Jiazhong

nese pottery and porcelain, and when he has time enjoys exploring the sites of historic old kilns in the Jingdezhen area. He is a council member of the Jingdezhen Artists Association and a member of the Jiangxi Provincial Artists Association.

Woodblock Fine Arts Reproductions

Xu Xinyou

veloped an interest in sculpture, in which he majored at the Sichuan Arts Academy. After graduation he was assigned to the Jingdezhen Porcelain Research Institute in Jiangxi province. Jingdezhen, for centuries the most famous of China's porcelain centers, is noted for its artistic figurines as well as for fine tableware.

Kang's style combines the rich legacy of Jingdezhen structure and decoration (including the classic Jingdezhen blue-and-white coloring) with his own gift of simple, expressive lines. His most common motifs are animals or figures from history and legend. His several hundred animal pieces are marvels of vivacity. Features may be exaggerated for effect, but the dynamic, flowing lines seem to render the essential reality of each creature. His superbly colored owls and yaks are considered among his most successful creations, and a series on China's ancient scientists won first prize at the Chinese Technological Arts Exhibition.

A sculptor in porcelain needs to have a great deal of technical knowledge. Kang is particularly skilled in the delicate business of applying different colors and glazes and then adjusting the heat of the kiln to produce exactly the effect he wants. He is also famed as a scholar and critic of ancient Chi-

XU XINYOU, 47, is a 30-year veteran worker of Beijing's Rong Bao Zhai Shop nationally noted for its fine woodblock reproductions of works of art. Xu is an expert in the printing process, the last of the three crucial stages in reproduction — the first two being tracing and carving or engraving.

His task involves exactly matching the ink and colors of the original works and then applying them to the woodblocks in just the right quantities. He has reproduced a great variety of works in both the *gongbi* (meticulous brushwork) and *xieyi* (expressionistic) styles, and believes that the biggest challenge is to bring out the characteristics and personality of the original artist in order to produce a faithful copy.

To improve their professional skills, the shop has always encouraged its workers to study art themselves and financed their lessons. Every Spring Festival it holds a special exhibition of workers' paintings and calligraphy. Xu has taken advantage of these opportunities to learn the basic techniques of sketching and Chinese ink-and-brush painting. He goes to Beijing parks every chance he gets in spring and summer when flowers are in blossom, for these are his favorite subjects.

Xu's wife works in the dyestuff processing workshop of Rong Bao Zhai, and it looks like it's becoming a family tradition for the couple's two children are now numbered among the apprentices their father has had under training for several years.

Clay Figures

Xu Gensheng

XU GENSHENG, 27, is a professional artist with the Huishan Clay Figure Shop in Wuxi, Jiangsu province. Sturdy, colorful figurines made from the dark, adhesive clay of the Huishan (Hui Hill) area have been famous for 400 years. Huishan is dotted with temples, and the throngs of people who once regularly came there for religious services and temple fairs helped popularize the local clay figures. Huishan's sprightly clay dolls with boldly exaggerated features are now known throughout the country, and many households in the area now produce them as a sideline.

Xu Gensheng grew up in the area, and was an apprentice to Li Renrong, the most famous clay art master of his generaton at the Huishan Clay Figure Shop. While retaining the traditional style, Xu works hard to revitalize the old forms and is constantly coming up with new ideas. Inspiration often comes from his frequent visits to temples and scenic spots, theatrical and dance performances and gymnastic competitions.

The vivid postures and natural, flowing lines of his clay sculptures have given him a reputation as an up-and-coming young artist, and his works have been exhibited in the Democratic People's Republic of Korea, Singapore, Malaysia, the Philippines, Mauritania and Australia. In 1982 he delighted Japanese audiences with demonstrations of his skill.

Jiajiang Fine Paper

SHI FULI, 46, is a Jiajiang paper maker from the county of the same name in Sichuan province. Made by hand of the bamboo that covers the county's hills, this paper is famous as a medium for Chinese ink-and-brush paintings. He comes from an old family of paper makers, and is head of his own small family paper mill. Actually, the demand

Shi Fuli (in front) and Gao Jun

for the county's most famous product is so great that nowadays most families do some paper making, and their workshops are scattered all along the bamboo groves which line the hilly streams.

The Chinese invented paper thousands of years ago, and the Jiajiang variety dates back to the Ming dynasty (1368-1644). The old methods are still followed. The bamboo fibers are first boiled to form a kind of pulp, which is then strained and pressed over a fine screen to dry. Sometimes the green outer layer is peeled off before the bamboo is boiled, producing the best quality pure, white paper that artists use for their finest work.

GAO JUN, 43, assisting Shi Fuli, developed an interest in traditional paper-making methods when he was young and learned the trade from veteran craftsmen. He is now an expert with the staff of the China Science and Technology Museum.

Brocade Weaving

YE YONGZHOU was born to a poor peasant family in Sichuan province 46 years ago. His father died when he was eleven, and the whole family moved to the city of Chengdu, where the young boy was apprenticed to a brocade weaver. This kind of weaving is very complicated and exacting, and it took Ye seven years to become a full-fledged weaver. Eventually he became a master of the art, and his work is renowned for its intricate patterns, colorful designs and the density and neatness of the raised patterns.

Now a veteran technician with the Chengdu Municipal Brocade Factory, whose products are exported to many foreign countries, Ye likes to play chess in his spare time. His wife and two daughters are also brocade weavers.

Ye Yongzhou

Qin Zelun

QIN ZELUN, 58, is a veteran with 40 years' experience. Working in partnership with Ye Yongzhou, Qin is stationed above and behind the warp threads. As Ye throws the weft shuttle between the warps, Qin as the "drawer" pulls cords which change the position of the warp threads in predetermined sequence to form the raised pattern on the brocade.

Known as one of the master brocade weavers in the country, Qin learned the craft from his elder brother. He is skilled in weaving over 30 kinds of brocade patterns, many of them his own designs. Customers love their bright, rich colors and distinctive styles.